SEMIGROUPS

Academic Press Rapid Manuscript Reproduction

Proceedings of the Monash University Conference on Semigroups
held at the Monash University, Clayton, Victoria, Australia, October, 1979.

SEMIGROUPS

edited by

T. E. HALL
P. R. JONES
G. B. PRESTON
Department of Mathematics
Monash University
Clayton, Victoria
Australia

ACADEMIC PRESS **1980**
A Subsidiary of Harcourt Brace Jovanovich, Publishers

New York London Toronto Sydney San Francisco

ACADEMIC PRESS, INC.
111 Fifth Avenue, New York, New York 10003

United Kingdom Edition published by
ACADEMIC PRESS, INC. (LONDON) LTD.
24/28 Oval Road, London NW1 7DX

Library of Congress Cataloging in Publication Data

Main entry under title:

Semigroups.

Papers presented at a conference held at Monash University, Australia, Oct. 27–30, 1979.
1. Semigroups—Addresses, essays, lectures.
I. Hall, Thomas Eric. II. Jones, P.R. III. Preston, G.B.
QA171.S522 512′.2 80-23748
ISBN 0-12-319450-4

PRINTED IN THE UNITED STATES OF AMERICA

80 81 82 83 9 8 7 6 5 4 3 2 1

CONTENTS

CONTRIBUTORS

Numbers in parentheses indicate the pages on which authors' contributions begin.

C. J. Ash (167) Department of Mathematics, Monash University, Clayton, Victoria, Australia 3168

C. Bailey (21) Department of Mathematics, The University of Auckland, Private Bag, Auckland, New Zealand

G. R. Baird (21) Department of Mathematics, The University of Auckland, Private Bag, New Zealand

G. T. Clarke (159) Department of Mathematics, Monash University, Clayton, Victoria, Australia 3168

R. J. H. Dawlings (121) Mathematical Institute, University of St. Andrews, North Haugh, St. Andrews, KY16 9SS, United Kingdom

P. D. Finch (239) Department of Mathematics, Monash University, Clayton, Victoria, Australia 3168

P. A. Grossman (57) Department of Mathematics, Monash University, Clayton, Victoria, Australia 3168

T. E. Hall (145) Department of Mathematics, Monash Unversity, Clayton, Victoria, Australia 3168

J. M. Howie (111) Mathematical Institute, University of St. Andrews, North Haugh, St. Andrews, KY16 9SS, United Kingdom

P. R. Jones (27) Department of Mathematics, Monash University, Clayton, Victoria, Australia 3168

H. Lausch (57) Department of Mathematics, Monash University, Clayton, Victoria, Australia 3168

John Meakin (67) Department of Mathematics and Statistics, The University of Nebraska-Lincoln, Lincoln, Nebraska 68588

D. B. McAlister (1) Department of Mathematical Sciences, Northern Illinois University, De Kalb, Illinois 60115

R. McFadden (177) Department of Mathematical Sciences, Northern Illinois University, De Kalb,Illinois 60115

W. D. Munn (207) Department of Mathematics, University of Glasgow, University Gardens, Glasgow, G12 8QW United Kingdom

G. B. Preston (225) Department of Mathematics, Monash University, Clayton, Victoria, Australia 3168

H. E. Scheiblich (191) Department of Mathematics, Computer Science, and Statistics, University of South Carolina, Columbia, South Carolina 29208

Takayuki Tamura (85) Department of Mathematics, University of California, Davis, California, 95616

P. G. Trotter (133) Department of Mathematics, University of Tasmania, Hobart, Tasmania, Australia 7001

D. C. Trueman (103) Department of Mathematics, Monash University, Clayton, Victoria, Australia 3168

M. Yamada (47) Department of Mathematics, Shimane University, Matsue, Shimane, Japan

PREFACE

This small conference was made possible principally through the generous financial support of Monash University and could not have taken place without the active support of the dean of its faculty of science, Professor J. M. Swan. I know I speak for all participants in registering our gratitude to him.

The four-day conference, from October 27 to October 30, 1979, of which these are the proceedings, was the culmination for most of the speakers of an extended session or workshop of four weeks of research and discussion spent at Monash. The effects of this extended conference will be seen in the large number of cross references to each other that appear in the papers offered.

Speakers for the conference were selected by the organising committee consisting of Dr. T. E. Hall, Dr. P. R. Jones, and me. Two or three speakers that we had hoped might attend could not come either because of other commitments or because our financial resources were not quite sufficient. But, otherwise, the list of speakers, and hence the topics chosen, must be taken as principally indicating the mathematical preferences of the members of the organising committee in current research in semigroup theory.

An exception to this, which we are pleased to have, is in the paper "The formal structure of observational procedures," by P. D. Finch, who asked the organisers whether they would like a paper showing the importance of semigroups in the analysis of the foundations of scientific thinking. In this paper, starting from a position that regards science as being provided with a set of instruments or "pointers" that provide observations or "readings" from which scientists draw their inferences, Finch gives various interpretations of scientific procedures, and the objects of scientific thought, in terms of properties of the free semigroup of "pointers" and the related free semigroups of possible "readings." A distinction is made between classical and nonclassical science, the latter being when any observations that are made can affect possible future observations. The analysis of classical science leads to the model of classical statistics, while the model for nonclassical science is shown to be that of a sequential computing machine, the theory of which is of course part of semigroup theory.

The organisers invited D. B. McAlister to give a survey talk about inverse semigroups. His paper, "A random ramble through inverse semigroups," gives his very interesting response to this request. He chose to explore the development and present position of the theory of P-semigroups within the theory of inverse semigroups. P-Semigroups were first introduced in "Zig-zag representations and inverse semigroups" [*Journal of Algebra* **32** (1974), pp. 178–206] by D. B. McAlister and R. McFadden, and, as McAlister says, the notion of a P-semigroup arose from

H. E. Scheiblich's construction of free inverse semigroups [*Proceedings of the American Mathematical Society* **38** (1973), pp. 1–7]. The idea of a P-semigroup was an attractive one: The elements of a P-semigroup are ordered pairs (e, g), say, where e belongs to a semilattice and g belongs to a group, and the product of such elements, involving a further ordered set, is reminiscent of that of a semidirect product. Coordinatization of elements, as in the Rees theorem on completely 0-simple semigroups, has been an illuminating tool in semigroup theory from its beginnings. McAlister and McFadden's new construction covered, in addition to free inverse semigroups, important other classes of inverse semigroups. An intrinsic characterization of P-semigroups followed quickly in McAlister's paper, "Groups, semilattices, and inverse semigroups, II" [*Transactions of the American Mathematical Society* **196** (1974), pp. 351–370], where it was shown, in particular, that any E-unitary inverse semigroup is a P-semigroup, and vice versa. Combined with another result of McAlister [same *Transactions* **192** (1974), pp. 227–244] that any inverse semigroup is an idempotent separating morphic image of an E-unitary inverse semigroup, it was now clear that P-semigroups played a central part in inverse semigroup theory. Many authors have written about P-semigroups since 1974. McAlister's paper draws together the threads connecting much of this work and points to new possibilities and problems to be solved.

It would be invidious to select some of the other papers in the proceedings for separate comment, for their subjects were not chosen at the specific invitation of the organising committee. Many of the papers contain more material surveying the present state of the game in the topic chosen than (unfortunately) is normal in most published research papers nowadays, and have a consequent additional interest. They also contain a rich offering of new and previously unpublished results. We commend them to our readers.

It remains to offer thanks. First, to the department of Mathematics at Monash University for its various support services for the conference supplied through the good offices of its chairman Professor P. D. Finch. Second, to Mrs. Joan Williams who undertook the greater part of the meticulous and daunting task of preparing this manuscript in a state ready for photographic reproduction for printing. Additional help with typing was provided by Ann-Marie Vandenberg, and the drawings were made by Mrs. Jean Sheldon, to whom our grateful thanks also go. Third, I thank my co-committee members Drs. Hall and Jones whom I left in the lurch to make all the final arrangements for the production of these proceedings while I went elsewhere on research leave.

Finally, on behalf of the organising committee, I thank Academic Press for their efficient and courteous cooperation in all aspects of the arrangements for producing these proceedings.

G. B. Preston

Paris, April 1980

A RANDOM RAMBLE THROUGH INVERSE SEMIGROUPS

D. B. McAlister

Department of Mathematical Sciences, Northern Illinois University,
De Kalb, Illinois, U.S.A., 60115.

Abstract

In this expository paper, we use a naive approach to the structure of inverse semigroups to motivate the introduction of P-semigroups and E-unitary inverse semigroups. A proof of the so-called P-theorem, due to W.D. Munn, is used to simplify some existing results on inverse subsemigroups of, and congruences on, E-unitary inverse semigroups.

1. Semidirect products and *P*–semigroups

Suppose that S is an inverse semigroup. Then the set E of idempotents forms a semilattice, under the natural partial order, with $e \wedge f = ef$. Around each idempotent e there is a group $H_e = \{a \in S: aa^{-1} = e = a^{-1}a\}$ so that, inside S we have a structure which can be pictured as follows.

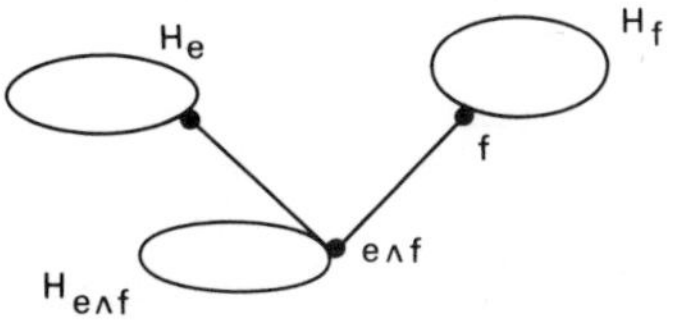

In general, of course, the groups H_e, $e \in E$ do not exhaust all of S and, indeed, it need not be true that $H_eH_f \subseteq H_{e\wedge f}$. A.H. Clifford [1941] showed, however, that if the idempotents of S are central then indeed

(i) $S = \cup\{H_e: e \in E\}$

(ii) $H_eH_f \subseteq H_{e\wedge f}$ for all $e,f \in E$

and, further, the multiplication in S is determined by a directed fami-

ISBN 0-12-319450-4

ly of homomorphisms $\phi_{e,f}: H_e \to H_f$, $e \geqslant f$, between the groups H_e, $e \in E$.

In this situation, the structure of S is completely determined in terms of groups and the semilattice E. In view of this, it is natural to wonder to what extent inverse semigroups can be constructed from groups and semilattices. Indeed much of the structure theory of inverse semigroups has been concerned with this problem. In this paper, I do not intend to describe the many deep fundamental contributions that have been made. Instead I shall take a naive approach by considering how one can put together groups and semilattices to get inverse semigroups.

Perhaps the most obvious way to combine a semilattice E and a group G is to form their direct product $E \times G$ under componentwise multiplication. This, certainly, gives an inverse semigroup. But, since its idempotents are the pairs $(e,1)$, $e \in E$, where 1 denotes the identity of G,

$$(e,1)(f,g) = (e \wedge f,g) = (f,g)(e,1),$$

for all $(f,g) \in E \times G$, so that idempotents are central. There are inverse semigroups which do not have central idempotents so we will not obtain all inverse semigroups in this fashion. One needs a mechanism to account for the non-centrality of idempotents. Such a mechanism is provided by the semidirect product construction.

DEFINITION 1.1. Let E be a semilattice and let G be a group which acts on E, on the left, by automorphisms. Then the semidirect product of E by G is the set of all pairs $(e,g) \in E \times G$ under the multiplication

$$(e,g)(f,h) = (e \wedge gf,gh).$$

The semidirect product of E by G is an inverse semigroup. The idempotents, as before, are the pairs $(e,1)$ with $e \in E$ but this time $(e,1)(f,h) = (e \wedge f,h)$ while $(f,h)(e,1) = (f \wedge he,1)$ so that, in general, idempotents are not central. The following proposition gives several elementary properties of semidirect products.

PROPOSITION 1.2. *Let E be a semilattice and let G be a group which acts on E, on the left, by automorphisms. Denote, by $P = P(G,E,E)$, the semidirect product of E by G. Then*

(i) the idempotents are the pairs $(e,1)$, $e \in E$; *they form a semilattice isomorphic to* E;

(ii) $(e,g)\mathcal{R}(f,h)$ *if and only if* $e = f$; $(e,g)\mathcal{L}(f,h)$ *if and only if* $g^{-1}e = h^{-1}f$;

(iii) the $\mathcal{H}$*-class corresponding to* $(e,1)$ *is isomorphic to the stabilizer of* e *under* G;

(iv) $(e,1)\mathcal{D}(f,1)$ *if and only if* $f = g.e$ *for some* $g \in G$; *thus the* $\mathcal{D}$*-classes of* P *correspond to the orbits of* E *under* G;

(v) $(e,1) \leq_{\mathcal{J}} (f,1)$ *if and only if* $g.e \leq f$ *for some* $g \in G$;

(vi) G *is isomorphic to the maximum group homomorphic image of* P *via the homomorphism* $(e,g) \to g$.

Unfortunately, one notices quickly that, if G is nontrivial, the semigroup $P = P(G,E,E)$ cannot have a zero so that even semidirect products on their own do not suffice if we hope to get all inverse semigroups. On this depressing note we pause to consider an example.

EXAMPLE 1.3. Let A_n be n-dimensional real affine space (say $n = 3$); then a geometric figure is a compact connected subset of A_n. The set F of geometric figures forms a semilattice under convex join on which the n-dimensional affine group G acts in an obvious way.

$$\alpha.A = A\alpha^{-1} \quad \text{for } A \in F,\ \alpha \in G.$$

This action gives rise to a semidirect product $P = P(G,F,F)$ of F by G.

It is interesting to look at the structure of P in terms of Green's relations, using proposition 1.2. Its idempotents are the pairs $(A,1)$ where $A \in F$ so these correspond to the geometric figures. Two idempotents $(A,1),(B,1)$ are $\mathcal{D}$-related if and only if $B = \alpha.A$ for some $\alpha \in G$; that is B and A are similar geometric figures. Likewise, $(A,1) \leq_{\mathcal{J}} (B,1)$ means that some translate of A is contained in B; that is, A can be shrunk and moved into B; that is, dim $A \leq$ dim B . Hence there is one $\mathcal{J}$-class for each dimension. One for points, one for finite line segments, one for plane figures etc. Note that $\mathcal{J} \neq \mathcal{D}$ so that there are principal factors which are 0-simple but not 0-bisimple.

This example shows that semidirect products of semilattices with groups can give rise to inverse semigroups with a sophisticated ideal structure so it is a satisfying example from an algebraic point of view but, from a naive geometric point of view it is not quite so satisfying. If one is working with geometric figures one wants to be able to get hold of them - one doesn't want them to be in the seventh galaxy beyond Andromeda. One way to accomplish this is to localize the figures by insisting that each of them contains the origin. Suppose we let E denote the set of all such objects and try a construction, like that in example 1, with E in place of F. An immediate problem arises! Because G contains linear translations, it moves things in E out of E so we can't quite construct a semidirect product of E by G. We want only to look at pairs (A,α) with $A \in E$ and $\alpha^{-1}A = A\alpha$ also in E. That is, we want to consider

$$\{(A,\alpha) \in E \times G\colon \alpha^{-1}A \in E\}.$$

This is an inverse subsemigroup of $P(G,F,F)$, which we shall write as $P(G,F,E)$. It is a special case of the construction described in definition 1.4.

DEFINITION 1.4. Let F be a (down) directed partially ordered set and let E be an ideal and subsemilattice of F. Let G be a group which acts on F by order automorphisms, on the left, in such a way that $F = G.E$ and set

$$P = P(G,F,E) = \{(e,g) \in E \times G\colon g^{-1}e \in E\}$$

under the multiplication

$$(e,g)(f,h) = (e \wedge gf, gh).$$

If G,E,F are as in definition 1.4, then $P(G,F,E)$ in an inverse semigroup with $(e,g)^{-1} = (g^{-1}e, g^{-1})$; we call it a P-semigroup. P-semigroups generalize semidirect products and provide more flexibility than is available with semidirect products alone. All of the properties of semidirect products which were described in proposition 1.2 hold for P-semigroups as well. However, as with semidirect products, P-semigroups in which the group is nontrivial, fail to have zeros so that not all inverse semigroups can be obtained in this way.

THEOREM 1.5. *Let S be an inverse semigroup. Then the following statements are equivalent:*

(i) S is isomorphic to a P-semigroup $P(G,F,E)$;

(ii) for all $e, a \in S$ *the equations* $ea = e = e^2$ *together imply* $a^2 = a$;

(iii) let σ *denote the minimum group congruence on* S; *then the idempotents of* S *form a* σ*-class;*

(iv) $\sigma \cap R = \Delta$, *the identical relation on* S.

The partially ordered set F in (i) need not be a semilattice. Necessary and sufficient conditions on S for F to be a semilattice are given in McAlister [1978b].

An inverse semigroup which obeys any of (ii), (iii), (iv) is called an E-unitary inverse semigroup. These semigroups were first introduced by Saîto [1965] and were then called proper inverse semigroups. Theorem 1.5 was originally proved by McAlister [1974b]. Since then it has been reproved several times; by Munn [1976d] and Schein [1975c] directly, and by Reilly and Munn [1976g] as a corollary to their investigations on congruences on P-semigroups. We shall indicate a proof, which is essentially that of Munn [1976d] following theorem 2.3.

The approach of Reilly and Munn [1976g] is interesting in that it makes use of the fact that the free inverse semigroup on a set can be regarded as a P-semigroup. Indeed it was the construction by Scheiblich [1973] of a free inverse semigroup which led to the notion of a P-semigroup and not the naive approach I have described above. The following result gives a slightly modified version of Scheiblich's characterization of the free inverse semigroup on a set X; it is similar to the construction in example 1 and is essentially that in McAlister and McFadden [1974c].

THEOREM 1.6. *Let* X *be a non-empty set and denote by* F_X *and* G_X, *respectively, the free semigroup and group on* X. *Then* G_X *is partially ordered by (we write* uw *to mean* uw *is reduced as it stands)*

$$u \leqslant v \quad \text{if and only if} \quad v = uw \quad \text{for some} \quad w \in F_X.$$

Let $\mathcal{X}$ *be the set of finite convex connected subsets of* G_X *and let* $\mathcal{Y} = \{A \in \mathcal{X}: 1 \in A\}$; $\mathcal{X}$ *is a semilattice under convex join and* $\mathcal{Y}$ *is an ideal of* $\mathcal{X}$. *Further* G_X *acts on* $\mathcal{X}$, *by automorphisms, under left multiplication. Then* $P(G_X, \mathcal{X}, \mathcal{Y})$ *is the free inverse monoid on* X.

Each member A of $\mathcal{Y}$ is a finite convex connected subset of G_X and so has a Haase diagram $H(A)$. This is a graph with vertices the set of elements of A and with a directed edge from u to v if v covers u in A; because A is convex, v covers u in A if and only if $v = ux$ for some $x \in X$. Further, since F_X is a free semigroup, and $\leqslant$ is defined by left division, it is immediate that $H(A)$ has no cycles; since it is connected, it is therefore a tree with 1 as a vertex.

The elements of $P = P(G_X, X, \mathcal{Y})$ are pairs $(A, w) \in \mathcal{Y} \times G_X$ with $w^{-1}A \in \mathcal{Y}$; that is, with $w \in A$. Hence, since A is determined by its Haase diagram $H(A)$, we can associate each element of P with a unique pair $(H(A), w)$ where $1, w$ are vertices of $H(A)$ and in this way we can obtain a graphical representation for the elements of P as trees with two distinguished vertices 1 and w. One can obtain another graphical representation by labelling the edges of $H(A)$ as follows: if v covers u, with $v = ux$ where $x \in X$, label the edge from u to v by x. Then the labelled graph $H(A)$, with the two distinguished vertices corresponding to 1 and w, is a birooted tree in the sense of Munn [1974d]. Conversely, if T is a birooted tree in the sense of Munn one can reverse the process to obtain T as the Haase diagram of a unique $A \in \mathcal{Y}$, by labelling the initial vertex of T by 1 and labelling other vertices using the labels of the shortest paths from 1, where b^{-1} is used to denote the edge labelled by b traversed backwards. The final vertex then corresponds to a unique $w \in A$ so that T corresponds to a unique $(A, w) \in P$. It follows from this that the representation, for the free inverse monoid on X, given in theorem 1.6 is exactly equivalent to that of Munn [1974b].

EXAMPLE 1.7. Let $X = \{x, y\}$. Then $A = \{1, x, y^{-1}, xy, xy^{-1}, y^{-1}x, y^{-2}, xyx, xy^2, xy^{-1}x, xy^{-1}x^{-1}\}$ belongs to $\mathcal{Y}$ so that $(A, xy^{-1}) \in P(G, X, \mathcal{Y})$. A has Haase diagram

$H(A)$

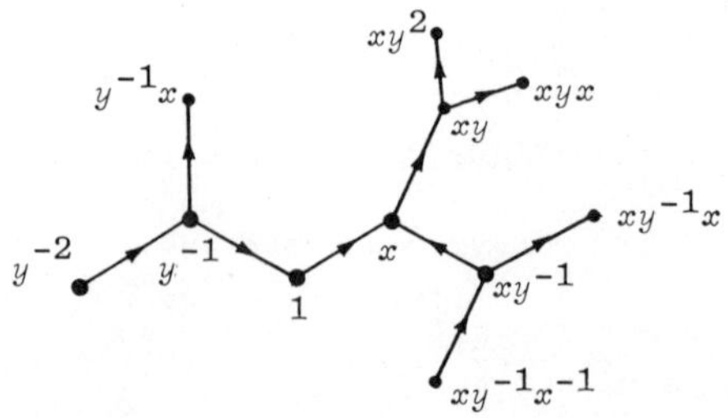

so that (A, xy^{-1}) gives rise to the birooted tree

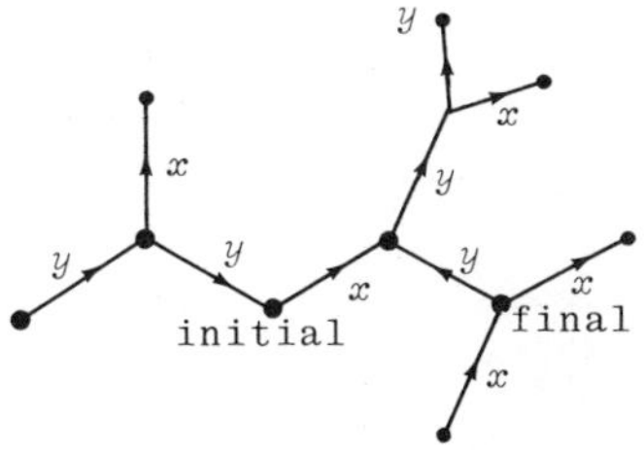

This birooted tree, in turn, gives rise to the pair (A, xy^{-1}).

Although not all inverse semigroups are E-unitary, many interesting inverse semigroups, in addition to free inverse semigroups, can be realized in this way. The following list describes some of these examples.

1. The class of E-unitary inverse semigroups is closed under products, free products and inverse subsemigroups.

That E-unitary inverse semigroups are closed under inverse subsemigroups and products is immediate. Closure under free products is shown in McAlister [1977a]. The class of E-unitary inverse semigroups is not closed under homomorphic images. (This follows because free inverse semigroups are E-unitary, but not all inverse semigroups are.)

2. Bisimple inverse monoids whose right unit subsemigroup is cancellative are E-unitary. Semilattices of groups whose connecting homomorphisms are one-to-one are E-unitary (McAlister [1974a]).

3. F-inverse semigroups (McFadden and O'Carroll [1971]). These can be characterized as P-semigroups $P(G,X,Y)$ with X a semilattice and Y a principal ideal of X; McAlister [1974b].

4. Projectives in the category of inverse semigroups; P.G. Trotter [A].

5. Translational hulls of E-unitary inverse semigroups; McAlister [1976b].

6. Simple inverse semigroups (not groups) whose lattice of full inverse subsemigroups is distributive; Jones [1978a].

7. Integrally closed Dubreil-Jacotin inverse semigroups; McFadden [1975a].

2. *E*–unitary covers for inverse semigroups

In their joint proof of theorem 1.5, Reilly and Munn [1976g] showed that every E-unitary congruence ρ on a P-semigroup has a P-semigroup as quotient. Thus, since free inverse semigroups can be realized as P-semigroups, so can all E-unitary inverse semigroups. They also show that if S is E-unitary and ρ is the smallest congruence on S inducing a given partition on the idempotents, then S/ρ is E-unitary. It follows that the following result is true; McAlister [1974a].

THEOREM 2.1. *Every inverse semigroup S is an idempotent separating homomorphic image of an E-unitary inverse semigroup P. We say that P is an E-unitary cover for S.*

EXAMPLE 2.2. Let G be a group and denote by $K(G)$ the set of all right cosets Ha of G modulo subgroups H of G. Then $K(G)$ becomes an inverse semigroup if we define

$$Ha * Kb = (H \vee aKa^{-1})ab;$$

this is the smallest coset of G which contains the set product $HaKb$ of Ha and Kb. The semigroup $K(G)$ is not E-unitary, unless G is trivial; however we can obtain it as an idempotent separating homomorphic image of an E-unitary inverse semigroup in the following way.

The group G acts on its $\vee$-semilattice L_G of subgroups by conjugation: $a.H = aHa^{-1}$. Thus we can form the semidirect product $P(G, L_G, L_G)$ of L_G by G. The mapping θ: $(H,a) \to Ha$ is then an idempotent separating homomorphism of $P(G, L_G, L_G)$ onto $K(G)$.

In example 2.2, the group G actually acts on the lattice L_G of subgroups by automorphisms so we can form the semidirect product of the $\wedge$-semilattice of subgroups of G by G. Its elements are pairs (A,a) with $A \leqslant G$ and $a \in G$ but, this time, the product is

$$(A,a) * (B,b) = (A \cap aBa^{-1}, ab).$$

In light of example 2.2, it is natural to wonder if there is another inverse semigroup which corresponds to $P(G, L_G, L_G)$, under this guise.

There is such a semigroup. For each $a \in G$ and subgroup A of G, conjugation by a defines an isomorphism of A onto $a^{-1}Aa$; we denote this map by $i_a|A$. The set of all such local inner isomorphisms of G is an inverse subsemigroup IA_G of the inverse semigroup A_G of isomorphisms between subgroups of G and the mapping $\theta: (A,a) \to i_a|A$ is an idempotent separating homomorphism of $P(G,L_G,L_G)$ onto IA_G.

According to theorem 2.1, there is a P-semigroup which has A_G as an idempotent separating homomorphic image. The discussion above shows that $P(G,L_G,L_G)$ will serve as such a P-semigroup if each member of A_G is a local inner isomorphism, and gives a clue to how the general procedure may be handled.

It is a theorem of Higman, Neumann and Neumann [1949] that each group G can be embedded in a group H in such a way that each isomorphism between subgroups of G is induced by conjugation by an element of H. Let L_G denote the $\wedge$-semilattice of subgroups of G and L_H that of H. Then L_G is an ideal of L_H so we may form the P-semigroup $P(H,L_H,L_G)$. Its elements consist of all pairs (A,a) with $A, a^{-1}A \in L_G$; that is, with A and $a^{-1}Aa$ subgroups of G; multiplication is defined by

$$(A,a)(B,b) = (A \cap aBa^{-1}, ab).$$

The map which sends (A,a) to $i_a|A$ is an idempotent separating homomorphism onto A_G.

What this example points out, I think, is that the search for E-unitary covers for inverse semigroups has a natural interpretation. It fits into the framework of isomorphism extension theorems like that of Higman and the Neumanns and thus, in turn, is related to questions about amalgamations.

When we turn away from specific examples, like the one we have just considered, more abstract arguments become necessary. Suppose that P is an E-unitary cover for an inverse semigroup S, and let G be the maximum group homomorphic image of P. Thus we have the following diagram of maps

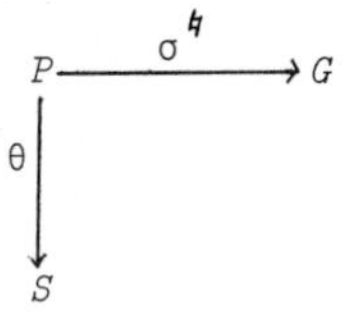

For each $s \in S$, set

$$s\phi = \{g \in G: s = p\theta,\ g = p\sigma^{\natural} \text{ for some } p \in P\} = s\theta^{-1}\sigma^{\natural}.$$

Then one can readily verify that $s\phi$ is a coset of G so that ϕ is actually a mapping of S into the inverse semigroup $K(G)$. Furthermore, for each $s,t \in S$, we have

$$(st)\phi \leqslant s\phi\ t\phi.$$

(Recall that the natural partial order on $K(G)$ is the reverse of inclusion.) That is, ϕ is a prehomomorphism of S into $K(G)$.

If $s\phi$ is an idempotent of $K(G)$ (i.e. a subgroup of G) then $1 \in s\phi$ so that $s = p\theta$ for some $p \in P$ with $p\sigma^{\natural} = 1$. Since P is E-unitary p must be idempotent and hence so is $s = p\theta$. Thus ϕ is idempotent determined in the sense that only idempotents are mapped to idempotents. In addition, since θ and $\sigma^{\natural}$ are both onto, $G = \cup\{s\phi: s \in S\}$.

On the other hand, let ϕ be an idempotent determined prehomomorphism of S into $K(G)$ such that $G = \cup\{s\phi: s \in S\}$ and set

$$P = P(\phi) = \{(s,g) \in S \times G: g \in s\phi\}.$$

THEOREM 2.3. *$P(\phi)$ is an E-unitary inverse semigroup.*

(i) The mapping $\theta: (s,g) \to s$ is an idempotent separating homomorphism of $P(\phi)$ onto S.

(ii) $P(\phi)$ has maximum group homomorphic image G under the mapping $\sigma^{\natural}: (s,g) \to g$.

Conversely, each E-unitary cover of S, with maximum group homomorphic image G, is isomorphic to $P(\phi)$ for some idempotent determined prehomomorphism ϕ of S into $K(G)$, such that $G = \cup\{s\phi: s \in S\}$.

EXAMPLE 2.4. Let S be any inverse semigroup and let ρ be an embedding of S into the symmetric inverse semigroup I_X on a set X. Let $Y = X$ if X is finite and, otherwise, let $Y = X \cup X'$ where $|X| = |X'|$ and $X \cap X' = \square$. For each $s \in S$, let

$$s\phi = \{\alpha \in S_Y: s\rho \subseteq \alpha\},$$

where S_Y denotes the symmetric group on Y, and let $G = \cup\{s\phi: s \in S\}$. Then G is a subgroup of S_Y and ϕ is an idempotent determined prehomomorphism of S into $K(G)$. Hence $P(\phi)$ is an E-unitary cover for S. This shows the truth of theorem 2.1.

Theorem 2.3 is due to McAlister and Reilly [1977b]. While it gives a simple description of an E-unitary cover in terms of ϕ, it does not give its structure as a P-semigroup. Two of the three components required for such a description are obvious; the group is G and the semilattice E is isomorphic to the semilattice E_S of idempotents of S. The third component F and the action of G on F can be obtained explicitly as follows.

Define a relation $\preccurlyeq$ on $G \times E_S$ by

$$(g,e) \preccurlyeq (h,f) \text{ if and only if there exists } s \in S \text{ such that}$$
$$ss^{-1} = e,\ s^{-1}s \leqslant f,\ g^{-1}h \in s\phi.$$

Then $\preccurlyeq$ is a quasiorder on $G \times E_S$ so that, if we let $\sim$ denote the equivalence relation associated with $\preccurlyeq$, $F = G \times E_S/\sim$ becomes a partially ordered set under

$$[g,e] \leqslant [h,f] \quad \text{if and only if} \quad (g,e) \preccurlyeq (h,f)$$

(where $[g,e]$ denotes the $\sim$-class of (g,e)). Indeed F is a down directed set with an ideal and subsemilattice

$$E = \{[1,e] : e \in E_S\} \approx E_S.$$

Further G acts on F by automorphisms as follows:

$$g.[h,f] = [gh,f].$$

Then $P(G,F,E) \approx P(\phi)$ and θ defined by

$$([1,e],g)\theta = s \quad \text{if} \quad ss^{-1} = e,\ g \in s\phi$$

is an idempotent separating homomorphism of $P(G,F,E)$ onto S.

If S is E-unitary with maximum group homomorphic image G under the homomorphism $\sigma^{\natural}$ then $\phi: S \to K(G)$ defined by

$$s\phi = \{s\sigma^{\natural}\}$$

is a prehomomorphism of S into $K(G)$ and the construction above reduces to Munn's construction [1976d] for S as a P-semigroup. It is determined by the quasiorder on $G \times E_S$ defined by

$(g,e) \prec (h,f)$ if and only if there exists $s \in S$ such that

$$ss^{-1} = e, \; s^{-1}s \leqslant f \text{ and } s\sigma^{\natural} = g^{-1}h.$$

3. Inverse subsemigroups and congruences.

Let $P = P(G,F,E)$ be a P-semigroup and let S be an inverse subsemigroup of P. Then S is an E-unitary inverse semigroup and so can be expressed as a P-semigroup. If we let

$$U = \{e \in E\colon (e,g) \in S \text{ for some } g \in G\}$$
$$H = \{g \in G\colon (e,g) \in S \text{ for some } e \in E\}$$

then U is a subsemilattice of E and H is a subgroup of G and is isomorphic to the maximum group homomorphic image of S. Hence S can be represented as a P-semigroup $P(H,K,U)$ where K is a down directed partially ordered set containing U as an ideal, on which H acts in such a way that $K = H.U$. In general, the action of H on K will not coincide with that induced from the action of G on F. Indeed, the subset HU of F need not be down directed with U as an ideal. Hence S need not be isomorphic to $P(H,HU,U)$ as a P-semigroup.

DEFINITION 3.1 [Jones 1976a]. Let S be an inverse subsemigroup of an inverse semigroup T. Then S is [weakly] partially unitary in T if and only if $tt^{-1} \in S$, $et \in S$, $[\,tt^{-1} \leqslant f$, for some $f^2 = f \in S]$, $e^2 = e \in T$ together imply $t \in S$.

If S is E-unitary the condition $et \in S$, in Definition 3.1, can be replaced by the hypothesis $t\sigma^{\natural} \in S\sigma^{\natural}$. Using this, Jones [1976a] proves the following result.

THEOREM 3.2. *Let $P(G,F,E)$ be an E-unitary inverse semigroup. Suppose that S is an inverse subsemigroup of S and let U,H be as above.*

(i) S is weakly unitary if and only if U is an ideal in HU and, for each $b \in HU$, there exists $u \in U$ such that $b \geqslant u$. In this case $S = P(H,HU,U)$.

(ii) S is partially unitary if and only if $E \cap HU = U$ and, for each $b \in HU$, there exists $u \in U$ such that $b \geqslant u$. In this case, U is an ideal of HU and $S = P(H,HU,U) = (U \times H) \cap P(G,F,E)$.

Any ideal of a $P(G,F,E)$ is partially unitary and, with the notation above U is an ideal of E while $H = G$. From this remark and theorem 3.2, one obtains O'Carroll's characterisation [1976e] of the ideals of $P(G,F,E)$ as the semigroups $P(G,GU,U)$ where U is an ideal of E such that $GU \cap E = U$.

Using the description of E-unitary inverse semigroups as P-semigroups, due to Munn [1976d], which was given at the end of Section 2, it is easy to give a representation for any inverse subsemigroup of $P(G,F,E)$.

PROPOSITION 3.3. *Let S be an inverse subsemigroup of $P(G,F,E)$ and let $U = \{e \in E\colon (e,g) \in S$ for some $g \in G\}$, $H = \{g \in G\colon (e,g) \in S$ for some $e \in E\}$. Define a quasiorder on $H \times U$ by*

$$(g,u) \preccurlyeq (h,v) \text{ if and only if } gu \leqslant hv \text{ and } (u,g^{-1}h) \in S,$$

and let $K = H \times U/\sim$ where $\sim$ is the equivalence associated with $\preccurlyeq$. Then K is a down directed set containing $\overline{U} = \{[1,u]\colon u \in U\}$ as an ideal isomorphic to U; $[1,u]$ denotes the $\sim$-class of $(1,u)$. Further H acts on K by $h.[g,u] = [hg,u]$ so we can form the P-semigroup $P(H,K,\overline{U})$. It is isomorphic to S under the mapping: $([1,u],h) \mapsto (u,h)$.

In the terminology of definition 3.1, S is weakly unitary if and only if the quasiorder in proposition 3.3 reduces to

$$(g,u) \leqslant (h,v) \text{ if and only if } gu \leqslant hv.$$

Thus K is isomorphic to HU under the mapping $gu \mapsto hv$ and we obtain the identification of $P(H,K,\overline{U})$ with $P(H,HU,U)$ described in theorem 3.2.

Jones [1976a] shows that, for each subgroup H of G and each subsemilattice U of E, the set of inverse subsemigroups S of $P(G,F,E)$, with

$$H = \{g \in G\colon (e,g) \in S \text{ for some } e \in E\},$$
$$U = \{e \in E\colon (e,g) \in S \text{ for some } g \in G\},$$

is a $\vee$-complete sublattice of the lattice of inverse subsemigroups of $P(G,F,E)$; it need not be complete. If (H,U) are compatible, in the sense that U is an ideal of HU and for each $b \in HU$ there exists $u \in U$ with $b \geqslant u$, then $P(H,HU,U)$ is the largest element of this sublattice.

In their joint proof of theorem 1.3, Reilly and Munn [1976g] make substantial use of a construction for the E-unitary quotients of a P-semigroup. This construction is based on the observation that every homomorphism on an E-unitary inverse semigroup S factors into an idempotent determined homomorphism and an idempotent separating homomorphism. Further the quotient of S by any idempotent determined congruence is again E-unitary. Thus the characterization of the E-unitary quotients of S breaks naturally into two parts: (i) the characterization of the idempotent determined quotients; (ii) the characterization of E-unitary idempotent separating quotients of P-semigroups.

THEOREM 3.4. *Let* $P = P(G,F,E)$ *be a* P*-semigroup and let* π *be an equivalence on* E *such that the following hold:*

(i) $(e,f) \in \pi$ *implies* $(u \wedge e,\ u \wedge f) \in \pi$ *for all* $u,e,f \in E$;

(ii) $(e,f) \in \pi$ *implies* $(g(a \wedge e),\ g(a \wedge f)) \in \pi$ *for all* $a \in E$, $g \in G$ *such that* $ga \in E$;

and define, for $(a,g),\ (b,h) \in P$,

$$(a,g)\pi^*(b,h) \text{ if and only if } a\ \pi\ b \text{ and } g = h.$$

Then π^* *is an idempotent determined congruence on* P *and each such has this form for a unique equivalence* π *on* E *satisfying (i) and (ii).*

Let $\mathcal{E}$ *denote the set of* π*-classes of* E *and define* $\preccurlyeq$ *on* $G \times \mathcal{E}$ *as follows:*

$$(g,A) \preccurlyeq (h,B) \text{ if and only if } ga \leqslant hb \text{ for some } a \in A,\ b \in B.$$

Then $\preccurlyeq$ *is a quasiorder on* $G \times \mathcal{E}$ *and* $\mathcal{X} = G \times \mathcal{E}/\sim$ *forms a down directed partially ordered set having an ideal and subsemilattice* $\mathcal{Y} = \{[1,A] : A \in \mathcal{E}\}$ *isomorphic to* $\mathcal{E}$; *as usual* $\sim$ *denotes the equivalence associated with* $\preccurlyeq$ *and* $[g,A]$ *denotes the class containing* (g,A). *Further* G *acts on* $\mathcal{X}$ *by order automorphisms as follows:* $h.[g,A] = (hg,A)$ *and* $S/\pi^* \approx P(G,\mathcal{X},\mathcal{Y})$.

Theorem 3.4 is a variant of a theorem of Reilly and Munn [1976g]. The first part including the description of π^* follows straight forwardly from the definition of the smallest congruence generated by a normal partition on the idempotents of an inverse semigroup; Reilly and Scheiblich [1967].

For the second part, let $T = S/\pi^*$. Then, since π^* is idempotent determined $T/\sigma^\natural \approx G$. To carry out Munn's construction of T as a P-semigroup, we define a quasiorder on $G \times E$ as follows:

$(g,A) \preccurlyeq (h,B)$ if and only if $tt^{-1} = A$, $t^{-1}t \leqslant B$ and $g^{-1}h = t\sigma^\natural$ for some $t \in T$.

Since $t = (u,k)\pi^*$ and $\pi^* \subseteq \sigma$, this implies $k = g^{-1}h$ and $u \in A$. Further, $k^{-1}u = h^{-1}gu\,\pi\,(h^{-1}gu \wedge b)$ for $b \in B$. Let $a = u \wedge g^{-1}hb$. Then $a\,\pi\,u$ and $(a,g^{-1}h)$ in S is such that $a \in A$, $h^{-1}ga = h^{-1}gu \wedge b \leqslant b$; thus $ga \leqslant hb$. Hence $(g,A) \preccurlyeq (h,B)$ implies $ga \leqslant hb$ for some $a \in A$, $b \in B$. Conversely, if $ga \leqslant hb$ with $a \in A$, $b \in B$, then $(a,g^{-1}h) \in S$. Let $t = (a,g^{-1}h)\pi^*$. Then $tt^{-1} = A$, $t^{-1}t \leqslant B$ and $g^{-1}h = t\sigma^\natural$ so that $(g,A) \preccurlyeq (h,B)$. It follows that T is determined, via Munn's construction, from the quasiorder

$(g,A) \preccurlyeq (h,B)$ if and only if $ga \leqslant hb$ for some $a \in A$, $b \in B$.

The determination of idempotent separating E-unitary congruences, and their quotients, on P-semigroups $P(G,F,E)$ is much less complicated to describe.

THEOREM 3.5 [Reilly and Munn, 1976g]. *Let $P = P(G,F,E)$ be a P-semigroup and let N be a normal subgroup of G such that*

$$N \subseteq \{g \in G\colon a = ga \text{ for all } a \in E \text{ with } g^{-1}a \in E\}$$

and define a relation ρ_N on P by

$(a,g)\rho_N(b,h)$ *if and only if* $a = b$ *and* $g^{-1}h \in N$.

Then ρ_N is an idempotent separating E-unitary congruence on P. Conversely each such has this form.

Define a relation κ on E by

$(a,b) \in \kappa$ *if and only if* $a = nb$ *for some* $n \in N$.

Then F/κ is a down directed partially ordered set under the partial order

$a\kappa \leqslant b\kappa$ *if and only if* $a \leqslant nb$ *for some* $n \in N$

and contains $E/\kappa = \{a\kappa\colon a \in E\}$ as an ideal isomorphic to E. Further G/N acts on F/κ by $Ng.A = gA$ for each $Ng \in G/N$, $A \in F/\kappa$. Then

$P/\rho_N \approx P(G/N, F/\kappa, E/\kappa)$.

Reilly and Munn [1976g] also show that the E-unitary congruences on a P-semigroup S form a complete lattice Λ^* under inclusion. However Λ^* need not be a sublattice of the lattice of all congruences on S. Let π be a normal partition on the idempotents of S and let α, β be E-unitary congruences on S which induce the partition π. Then Reilly and Munn also show that $\{\gamma \in \Lambda^*: \alpha \subseteq \gamma \subseteq \beta\}$ is a complete modular sublattice of Λ^*.

Other results on congruences, and the lattice of congruences, on E-unitary semigroups have been obtained by Jones [1976a] and McAlister [1974b, 1976c].

4. Idempotent determined congruences on inverse semigroups

E-unitary inverse semigroups can be characterised as those inverse semigroups which admit an idempotent determined homomorphism onto a group. In this light, theorem 1.3 can be regarded as a construction for an extension of a group by an idempotent determined congruence. O'Carroll [1975b, 1976e,f, 1977c] has generalized theorem 1.5 to give a construction for extensions of inverse semigroups by idempotent determined congruences. The ingredients involved are similar to those involved in the construction of P-semigroups.

Let X be a down directed partially ordered set with Y an ideal and subsemilattice of X. Let T be an inverse semigroup with semilattice of idempotents E and let θ be a homomorphism of Y onto E. Then T is said to act effectively on X if there is an anti-homomorphism of T into the inverse semigroup of isomorphisms between non-empty ideals of X such that the following hold:

(i) for each $a \in Y$ and $e^2 = e \in T$,
$$a \in \Delta e \text{ if and only if } a\theta \leqslant e$$

(ii) $X = T.Y$

(iii) for each $t \in T$ there exists $a \in Y$ with $a\theta = tt^{-1}$ and $t^{-1}a \in Y$.

In these conditions we preserve the analogy with group actions by writing ta for the image of a under the partial isomorphism of X corresponding to $t \in T$ whenever a belongs to the domain Δt.

THEOREM 4.1 [O'Carroll]. *Let T, X, Y, θ be as above and set*

$$L_m = L_m(T,X,Y) = \{(a,t) \in Y \times T: tt^{-1} = a\theta,\ t^{-1}a \in Y\}$$

under the multiplication

$$(a,t)(b,s) = (t(t^{-1}a \wedge b), st).$$

Then L_m is an inverse semigroup with semilattice of idempotents isomorphic to Y and θ defined by $(a,t)\theta = t$ is an idempotent determined homomorphism of L_m onto T.

Conversely, let S be an inverse semigroup with semilattice of idempotents isomorphic to Y and let θ be an idempotent determined homomorphism of S onto an inverse semigroup T. Then $S \approx L_m(T,X,Y)$ for some effective action of T on X.

O'Carroll has given two proofs of this result, similar to those of McAlister [1974b] and Schein [1975c] for P-semigroups, in O'Carroll [1975b, 1976e] together with some refinements in O'Carroll [1977c]. As was the case with theorem 1.3, theorem 4.1 admits yet another proof; one based on the ideas in Munn [1976d]. For completeness we sketch the construction process of $L_m(T,X,Y)$ from an idempotent determined homomorphism θ of S onto T using this approach.

Let E denote the semilattice of idempotents of S and set $X = \{(t,a) \in T \times E: a\theta = t^{-1}t\}$ and define

$$(t,a) \preccurlyeq (u,b) \Longleftrightarrow tt^{-1} \leqslant uu^{-1} \text{ and } ss^{-1} = a,\ s^{-1}s \leqslant b \text{ and } s\theta = t^{-1}u,$$

for some $s \in S$.

Then $\preccurlyeq$ is a quasiorder on X and $X = X/\sim$ is a down directed partially ordered set containing $Y = \{[a\theta, a]: a^2 = a \in S\}$ as an ideal isomorphic to the semilattice of idempotents of S. (Here as usual $\sim$ denotes the equivalence corresponding to $\preccurlyeq$ and $[t,a]$ denotes the $\sim$-class of (t,a).)

For each $t \in T$, let $\Delta t = \{[u,b] \in X: uu^{-1} \leqslant t^{-1}t\}$ and set $t.[u,b] = [tu,b]$ for each $[u,b] \in \Delta t$. Then this defines an effective action of T on X and $S \approx L_m(T,X,Y)$.

Theorem 4.1 provides a mechanism for constructing other classes of inverse semigroups, than E-unitary semigroups, by replacing P-semigroups by the semigroups $L_m(T,X,Y)$. O'Carroll [1977c] has characterized those semigroups of the form $L_m(T,X,Y)$ with T a semilattice of groups.

THEOREM 4.2. *Let S be an inverse semigroup. Then the following are equivalent:*

(1) S is a semilattice of E-unitary inverse semigroups;

(2) for all $x,y \in S$, $e \in E^1$, $xye \in E$ implies $yxe \in E$, where E denotes the set of idempotents of S; that is, S is strongly E-reflexive;

(3) $S \approx L_m(T,X,Y)$ for some T,X,Y with T a semilattice of groups.

Theorem 4.2 shows that strongly E-reflexive inverse semigroups are the semigroups obtained if the naive approach described in Section 1 is followed when the group involved is allowed to vary with the idempotents. This procedure gives perhaps a more realistic model for the structure of inverse semigroups than is provided by the construction of P-semigroups. However it can be shown that the inverse semigroup of 2×2 matrix units is not strongly E-reflexive so that theorem 4.2 is not generally applicable.

References

1941 A.H. Clifford, Semigroups admitting relative inverses. *Annals of Math.*, 42 (1941), 1037-1049.

1949 G. Higman, B.H. Neumann and H. Neumann, Embedding theorems for groups. *J. London Math. Soc.*, 24 (1949), 247-254.

1965 T. Saîto, Proper ordered inverse semigroups. *Pac. J. Math.*, 15 (1965), 649-666.

1967 N.R. Reilly and H.E. Scheiblich, Congruences on regular semigroups. *Pac. J. Math.*, 23 (1967), 349-360.

1971 R. McFadden and L. O'Carroll, F-inverse semigroups. *Proc. London Math. Soc.* (3), 22 (1971), 652-666.

1973 H.E. Scheiblich, Free inverse semigroups. *Proc. Amer. Math. Soc.*, 38 (1973), 1-7.

1974a D.B. McAlister, Groups, semilattices and inverse semigroups. *Trans. Amer. Math. Soc.*, 192 (1974), 227-244.

1974b D.B. McAlister, Groups, semilattices and inverse semigroups, II. *Trans. Amer. Math. Soc.*, 196 (1974), 251-270.

1974c D.B. McAlister and R. McFadden, Zig-zag representations and inverse semigroups. *J. Algebra*, 32 (1974), 178-206.

1974d W.D. Munn, Free inverse semigroups. *Proc. London Math. Soc.* (3), 29 (1974), 385-404.

1974e L. O'Carroll, Reduced inverse and partially ordered semigroups. *J. London Math. Soc.* (2), 9 (1974), 293-301.

1975a R. McFadden, Proper Dubreil-Jacôtin inverse semigroups. *Glasgow Math. J.*, 16 (1975), 40-51.

1975b L. O'Carroll, Inverse semigroups as extensions of semilattices. *Glasgow Math. J.*, 16 (1975), 12-21.

1975c B.M. Schein, A new proof of the McAlister *P*-theorem. *Semigroup Forum*, 10 (1975), 185-188.

1976a P.R. Jones, The lattice of inverse subsemigroups of a reduced inverse semigroups. *Glasgow Math.J.*, 17 (1976), 161-172.

1976b D.B. McAlister, Some covering and embedding theorems for inverse semigroups. *J. Austral. Math. Soc.*, 22A (1976), 188-211.

1976c D.B. McAlister, v-prehomomorphisms on inverse semigroups. *Pac. J. Math.*, 67 (1976), 215-231.

1976d W.D. Munn, A note on *E*-unitary inverse semigroups. *Bull. London Math. Soc.*, 8 (1976), 71-76.

1976e L. O'Carroll, Idempotent determined congruences on inverse semigroups. *Semigroup Forum*, 12 (1976), 233-244.

1976f L. O'Carroll, Embedding theorems for proper inverse semigroups. *J. Algebra*, 42 (1976), 26-40.

1976g N.R. Reilly and W.D. Munn, *E*-unitary congruences on inverse semigroups. *Glasgow Math. J.*, 17 (1976), 57-75.

1977a D.B. McAlister, Inverse semigroups generated by a pair of subgroups. *Proc. Royal Soc. Edin.*, 77A (1977), 9-22.

1977b D.B. McAlister and N.R. Reilly, *E*-unitary covers for inverse semigroups. *Pac. J. Math.*, 68 (1977), 161-174.

1977c L. O'Carroll, Strongly *E*-reflexive inverse semigroups. *Proc. Edin. Math. Soc.*(2), 20 (1977), 339-354.

1978a P.R. Jones, Distributive inverse semigroups. *J. London Math. Soc.* (2), 17 (1978), 457-466.

1978b D.B. McAlister, *E*-unitary inverse semigroups over semilattices. *Glasgow Math. J.*, 19 (1978), 1-12.

NOT YET PUBLISHED

[A] P.G. Trotter, Projectives in inverse semigroups.

DIVISIBILITY IN CATEGORIES

C. Bailey and G. R. Baird

**Department of Mathematics, The University of Auckland,
Private Bag, Auckland, New Zealand.**

Abstract

In this paper we investigate divisibility in categories. The work is motivated by the well-known characterization of divisibility in the category of sets in terms of images and relations. To provide an appropriate framework for our discussion we introduce the notions of a category with images and a category with relations.

The purpose of this note is to extend some results of Fitz-Gerald [1973] on divisibility in categories with images.

We begin by recalling the following well known characterization of divisibility in the category of sets, due originally to G.B. Preston [1959].

THEOREM 1. *Let x and y be arrows in the category of sets.*

(i) If x and y have the same codomain then there exists an arrow z such that $x = zy$ if and only if $Im(x) \subseteq Im(y)$.

(ii) If x and y have the same domain then there exists an arrow z such that $x = yz$ if and only if $y \circ y^{-1} \subseteq x \circ x^{-1}$.

We seek appropriate generalizations of this result to other categories.

1. Categories with images

Let C be a category. We define a preorder on the collection of monos of C by saying that $m_1 \leqslant m_2$ if there exists an arrow n in C such

ISBN 0-12-319450-4

that $m_1 = nm_2$. We denote the equivalence class containing m_1 under the equivalence associated with $\leqslant$ by $[m_1]$ and, abusing language, denote also by $\leqslant$ the partial order induced on the equivalence classes of monos by $\leqslant$. Clearly, $[m_1] = [m_2]$ if and only if there exists an iso n in C such that $m_1 = nm_2$.

For arrows x,y and m of C, $x = ym$ is said to be an *image factorization* of x if (i) m is a mono and (ii) any mono right dividing x also right divides m. In this case m is said to be an *image* of x and y a *co-image* of x.

The category C is said to have *images* if every arrow of C has an image factorization. If $x = ym$ is an image factorization of an arrow x we define $Im(x) = [m]$.

Dually, we have the notions of *coimage factorization*, *coimage*, *co-coimage* and $Coim(x)$.

LEMMA 1. *Let $x = ym$ be an image factorization of an arrow x in C and 1 the identity arrow on the codomain of y. Then $y = y1$ is an image factorization of y.*

Lemma 2 (Fitz-Gerald [1973]). *Let C be a category with images. If $x = zy$ in C then $Im(x) \leqslant Im(y)$.*

The proofs of lemmas 1 and 2 are straightforward and are omitted.

THEOREM 2. *In a category C with images the following are equivalent.*

(i) For all arrows x and y in C, $Im(x) \leqslant Im(y)$ if and only if there exists an arrow z in C such that $x = zy$.

(ii) All co-images in C have left inverses.

Proof. We begin by showing that (i) implies (ii). Suppose that the arrow u is a co-image in C, that is there exists an arrow x in C with an image factorization $x = uy$. Let 1 be the identity arrow on the codomain of u. By lemma 1 $Im(u) = Im(1)$ and so $Im(1) \leqslant Im(u)$. Hence there exists an arrow z such that $1 = zu$ and u has a left inverse.

On the other hand suppose that co-images have left inverses. In view of lemma 2 we only have to prove that 'only if' part of (i). Suppose x and y are arrows in C such that $Im(x) \leqslant Im(y)$ and let $x = um$

and $y = vn$ be image factorizations. $Im(x) \leqslant Im(y)$ implies that $m \leqslant n$ and so there exists an arrow w such that $m = wn$. Let v^* be the left inverse of the co-image v and put $z = uwv^*$. Then $zy = (uwv^*)(vn) = x$.

THEOREM 3. *In a category with coimages the following are equivalent.*

(i) For all arrows x *and* y *in* C, $Coim(x) \leqslant Coim(y)$ *if and only if there exists an arrow* z *in* C *such that* $x = yz$.

(ii) All co-coimages in C *have right inverses.*

Theorem 3 follows from theorem 2 by duality.

2. Categories with relations

Let C be a category with products. A *relation* from an object A of C to an object B of C is a mono m with codomain $A \times B$; thus m is determined by arrows m_1 and m_2 of C with common domain and codomains A and B respectively. We shall write $m = \langle m_1, m_2 \rangle$.

An arrow x in C with domain A and codomain B defines the relation $x_1 = \langle 1_A, x \rangle$ from A to B, where 1_A is the identity arrow on A.

In a category with products, pullbacks and images one can always define the composition of two relations. For our purposes, however, it suffices, given an arrow x of C, to define $x_1 \circ x_1^{-1}$, the composition of x_1 with its inverse relation x_1^{-1}.

Let C be a category with products and pullbacks and x an arrow of C with domain A and codomain B. To define $x_1 \circ x_1^{-1}$ we form the pullback of x with itself

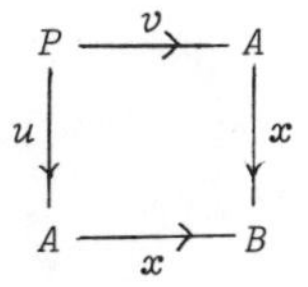

and put $x_1 \circ x_1^{-1} = \langle u, v \rangle$.

Of course, in the category of sets it is readily checked that $x_1 \circ x_1^{-1}$ is the usual composition of x with its inverse relation x^{-1}.

LEMMA 3. *Let C be a category with products and pullbacks. If $x = yz$ in C then $y_1 \circ y_1^{-1} \leqslant x_1 \circ x_1^{-1}$.*

LEMMA 4. *Let C be a category with products and pullbacks. If for all arrows x and y in C, $y_1 \circ y_1^{-1} \leqslant x_1 \circ x_1^{-1}$ implies that there exists z in C such that $x = yz$ then all monos in C are right invertible.*

The proofs of lemmas 3 and 4 are straightforward and are omitted. To obtain a converse to lemma 4 we worked within a topos.

A category C is called a topos if (i) C has all finite limits, (ii) C is cartesian closed and (iii) C has a subobject classifier. For a definition of these terms the reader is referred to Johnstone [1977].

The following lemma is extracted from chapter 1.5 of Johnstone [1977].

LEMMA 5. *In a topos image factorizations are coimage factorizations. Furthermore, a coimage of an arrow x is obtained by taking the coequalizer of the kernel pair of x.*

THEOREM 4. *Let x and y be arrows of a topos. Then $x_1 \circ x_1^{-1} \leqslant y_1 \circ y_1^{-1}$ if and only if $Coim(y) \leqslant Coim(x)$.*

Proof. If x and y do not have the same domain we have nothing to prove. Hence we assume that the domain of x is the domain of y. Let $x = em$ and $y = fn$ be coimage (= image) factorizations. Consider the following pullback diagrams.

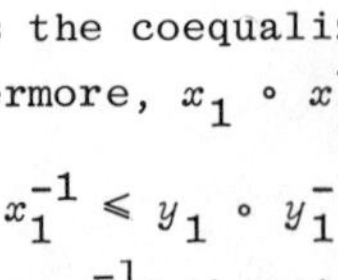

(a)

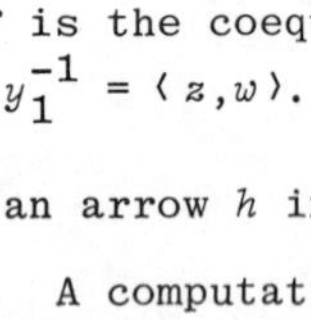

(b)

Now e is the coequalizer of u and v and f is the coequalizer of z and w; furthermore, $x_1 \circ x_1^{-1} = \langle u,v \rangle$ and $y_1 \circ y_1^{-1} = \langle z,w \rangle$.

If $x_1 \circ x_1^{-1} \leqslant y_1 \circ y_1^{-1}$ then there exists an arrow h in C such that $x_1 \circ x_1^{-1} = h(y_1 \circ y_1^{-1})$, that is $u = hz$ and $v = hw$. A computation shows that $vf = uf$ and since e is the coequalizer of v and u there exists an arrow k in e such that $f = ek$. Thus $f \leqslant e$ and $Coim(y) \leqslant Coim(x)$.

Conversely, suppose that $Coim(y) \leqslant Coim(x)$, that is there exists an arrow k in C such that $f = ek$. A computation shows that $vy = uy$. The pullback diagram (b) provides an arrow h of C such that $v = hw$ and $u = hz$. Thus $\langle v,u \rangle = h\langle z,w \rangle$ and so $x_1 \circ x_1^{-1} \leqslant y_1 \circ y_1^{-1}$.

COROLLARY 1. *In a topos C the following are equivalent.*

(i) For all arrows x and y in C, $y_1 \circ y_1^{-1} \leqslant x_1 \circ x_1^{-1}$ *if and only if there exists an arrow z in C such that* $x = yz$.

(ii) All monos in C have right inverses.

Corollary 1 follows from theorems 3 and 4 and lemmas 3 and 4.

References

1959 G.B. Preston, Embedding any semigroup in a *D*-simple semigroup. *Trans. Amer. Math. Soc.* 93 (1959), 351-355.

1973 D.G. Fitz-Gerald, *Divisibility in the Category of Binary Relations and Kindred Categories.* Ph.D. thesis, Monash University, 1973.

1977 P.T. Johnstone, *Topos Theory.* Academic Press, London, 1977.

UNIVERSAL ASPECTS OF COMPLETELY SIMPLE SEMIGROUPS

P. R. Jones

Department of Mathematics, Monash University, Clayton,
Victoria, Australia, 3168.

Abstract

The class $\underset{\sim}{CS}$ of completely simple semigroups forms a variety when considered as semigroups with the additional unary operation $x \mapsto x^{-1}$, x^{-1} being the inverse of x in the subgroup H_x. We present a Rees matrix representation for the $\underset{\sim}{CS}$-free product of any family of completely simple semigroups and for the $\underset{\sim}{CS}$-free product of any amalgam of completely simple semigroups. From the former, a description of the free semigroups in $\underset{\sim}{CS}$ and of certain free idempotent-generated semigroups in $\underset{\sim}{CS}$ is easily deduced. We then consider the lattice of subvarieties of $\underset{\sim}{CS}$. Also discussed are injectives and projectives in $\underset{\sim}{CS}$ and a problem concerning the Hopf property.

1. Introduction

As examples of completely 0-simple semigroups, completely simple semigroups were studied in the earliest papers on semigroups. The study of the *class* $\underset{\sim}{CS}$ of completely simple semigroups began simultaneously, with the determination by Rees [1940] of the isomorphisms between completely [0-] simple semigroups. Later, various authors considered the lattice of congruences on completely [0-] simple semigroups (for a history, see [Howie 1976, III.4]). In [1968] McAlister showed that for any non-empty set X there is a *free completely simple semigroup* over X, which we denote here by $F_X(\underset{\sim}{CS})$. A stimulus to the study of completely simple semigroups as a class was provided by Petrich [1975, 1977e], who considered the class $\underset{\sim}{CR}$ of completely regular semigroups (unions of groups) as a *variety*, not of semigroups but

ISBN 0-12-319450-4

of semigroups with the additional unary operation $x \mapsto x^{-1}$, x^{-1} being the inverse of x in the subgroup H_x. Clearly the morphisms and congruences in this variety are just the semigroup morphisms and congruences. Now the class $\underset{\sim}{CS}$ becomes a subvariety of $\underset{\sim}{CR}$ and the semigroups $F_X(\underset{\sim}{CS})$ are just the free objects in the variety $\underset{\sim}{CS}$. Their structure has been determined by Clifford [1979a].

As a variety we may ask of $\underset{\sim}{CS}$ the usual questions asked of varieties of algebras in general. For instance it is known that the congruence lattice $\Lambda(S)$ of any completely simple semigroup is semimodular [Lallement 1967a]; however since the congruence lattice of a right zero semigroup S is just the full partition lattice $\Pi(S)$ on the set S, and since *any* lattice can be embedded in $\Pi(X)$ for some set X [Whitman 1946], there is no non-trivial lattice identity (that is, no identity not satisfied by all lattices) satisfied by $\Lambda(S)$ for *every* $S \in \underset{\sim}{CS}$. Similarly the lattice $L(S)$ of completely simple subsemigroups of S satisfies no non-trivial lattice identity for every $S \in \underset{\sim}{CS}$. The problem of actually *characterizing* those lattices which appear as $\Lambda(S)$, or as $L(S)$, for some $S \in \underset{\sim}{CS}$ is rather more difficult and, as far as we know, unsolved.

Turning now to free products of arbitrary families of completely simple semigroups and of arbitrary *amalgams* of completely simple semigroups we see that these naturally exist, since $\underset{\sim}{CS}$ is a variety. We provide representations of these products as Rees matrix semigroups (§§2,3). Clarke [A] shows that any amalgam of completely simple semigroups is strongly embeddable in a member of $\underset{\sim}{CS}$, that is $\underset{\sim}{CS}$ has the *strong amalgamation property*. He also considers the problem of finding which subvarieties of CS have this property.

Since $F_X(\underset{\sim}{CS})$ is just the $\underset{\sim}{CS}$-free product of a family of copies of $F_{\{x\}}(\underset{\sim}{CS})$, $x \in X$, (each an infinite cyclic group), the description given by Clifford [1979a] is easily obtained, together with a description of the *free idempotent generated completely simple semigroup* $T(\alpha,\beta)$ *with* β *L-classes and* α *K-classes* whose existence was discovered by Eberhart, Kinch and Williams [1973a] and whose structure was found by Pastijn [1977b].

Injectives in various classes of semigroups have been treated comprehensively by Schein [1974] and others. Recently interest in *projectives* in various classes of semigroups has been shown by Nordahl and

Scheiblich [G] and Trotter [I,J]. In §5 we use the above results to describe the injectives and projectives in $\mathcal{CS}$.

Petrich [1977c] posed the problem of describing the *lattice* $V(\mathcal{CS})$ *of varieties of completely simple semigroups*. We show (§6) that $V(\mathcal{CS})$ is modular (in fact more generally the lattice of varieties of bands of groups is modular [Hall and Jones, B]). We also decompose $V(\mathcal{CS})$ into interval sublattices: each interval consists of all varieties whose intersection with the variety of all groups is a fixed variety G, say, of groups; within the interval each variety is determined by a subgroup, invariant in a certain sense, of the $\mathcal{G}$-free group on a countably infinite set. Whilst not providing a "list" of identities in $\mathcal{CS}$, this decomposition (theorem 6.5) does enable us to deduce, for instance, the description of all subvarieties of $\mathcal{CS}$ with abelian subgroups, first given by Rasin [1979b]. It also yields a description of the *relatively* free completely simple semigroups. This topic is pursued further in [Jones, C].

Finally (§7) we show that for a finite set X, $F_X(\mathcal{CS})$ is Hopfian (that is, each of its onto endomorphisms is an automorphism) and consider the problem of whether $F_X(\mathcal{V})$ is Hopfian for *every* subvariety V of $\mathcal{CS}$.

At this point it is worthwhile remarking that most of the results we obtain for $\mathcal{CS}$ have direct analogues for the varieties $\mathcal{CS}(\mathcal{G})$, consisting of all completely simple semigroups whose subgroups belong to the given variety $\mathcal{G}$ of groups.

Various parts of this paper have been submitted for publication, in extended form, in [Jones, C,D].

2. Free products

If $\{U_\gamma : \gamma \in \Gamma\}$ is a family of completely simple semigroups, denote by $T = *_{\mathcal{CS}}\{U_\gamma : \gamma \in \Gamma\}$ their free product in $\mathcal{CS}$, defined in terms of the usual universality properties. That is, there exist morphisms $\iota_\gamma : U_\gamma \to T$, $\gamma \in \Gamma$, such that for any family of morphisms $\theta_\gamma : U_\gamma \to W$, $\gamma \in \Gamma$, for some $W \in \mathcal{CS}$, there is a morphism $\theta : T \to W$ such that $\iota_\gamma\theta = \theta_\gamma$ for all $\gamma \in \Gamma$. If for each $\gamma \in \Gamma$ we put $V_\gamma = U_\gamma\iota_\gamma$ then T is generated by $\cup\{V_\gamma : \gamma \in \Gamma\}$. In fact the ι_γ, $\gamma \in \Gamma$, are *embeddings*, since the family $\{U_\gamma : \gamma \in \Gamma\}$ can be *jointly embedded* in their direct product.

For each $\gamma \in \Gamma$ we may assume $U_\gamma = M(G_\gamma;I_\gamma,\Lambda_\gamma;P^{(\gamma)})$, where $P^{(\gamma)} = (p^{(\gamma)}_{\lambda i})$ is a $\Lambda_\gamma \times I_\gamma$ matrix, over the group G_γ, which we may further assume to be normalized with respect to some element of I_γ and some element of Λ_γ, both of which we shall denote by γ. Thus if e denotes the identity of G_γ,

$$p^{(\gamma)}_{\gamma i} = p^{(\gamma)}_{\lambda\gamma} = e \quad \text{for each } (\lambda,i) \in \Lambda_\gamma \times I_\gamma.$$

We may suppose also that $T = M(H;J,M;Q)$ for some $M \times J$ matrix Q, over the group H. Now each embedding ι_γ clearly induces bijections $i \to i'$, of I_γ into J, and $\lambda \to \lambda'$, Λ_γ into M such that $H_{i\lambda}\iota_\gamma \subseteq H_{i'\lambda'}$ for all $(i,\lambda) \in I_\gamma \times \Lambda_\gamma$. Let

$$J_\gamma = \{i' : i \in I_\gamma\} \quad \text{and} \quad M_\gamma = \{\lambda' : \lambda \in \Lambda_\gamma\}.$$

Consider Green's relation $\mathcal{H}$ on T. Since on any completely simple semigroup $\mathcal{H}$ is a congruence, in fact the least rectangular band congruence, and since rectangular bands form a subvariety $\underset{\sim}{RB}$ of $\underset{\sim}{CS}$, we have from universality

$$T/\mathcal{H} \simeq \mathop{*}_{\underset{\sim}{RB}}\{V_\gamma/\mathcal{H} : \gamma \in \Gamma\}.$$

Clearly the $\underset{\sim}{RB}$-free product of a family of $J_\gamma \times M_\gamma$-rectangular bands is just the $\dot{\cup}J_\gamma \times \dot{\cup}M_\gamma$-rectangular band. Thus J and M are the disjoint unions of the sets J_γ, $\gamma \in \Gamma$, and M_γ, $\gamma \in \Gamma$, respectively.

The crucial result in finding H and Q is the following normalization. First choose, arbitrarily, some member 0 of Γ.

LEMMA 2.1. *We may normalize Q so that for all $\gamma \in \Gamma$,*

(i) $q_{0'\gamma'} = e$,

(ii) $q_{\mu\gamma'} = e$ *for all* $\mu \in M_\gamma$ *(1)*

and (iii) $q_{\gamma' j} = e$ *for all* $j \in J_\gamma$.

Proof. The normalization corresponds (see lemma 3.6 of [Clifford and Preston 1961]) to the transformation

$$q_{\mu j} \to v_\mu q_{\mu j} u_j, \ (\mu,j) \in M \times J,$$

where if $\mu \in M_\tau$, $v_\mu = q_{0'\tau'} q^{-1}_{\mu\tau'}$

and if $j \in J_\delta$, $u_j = q^{-1}_{\delta' j} q_{\delta'\delta'} q^{-1}_{0'\delta'}$.

From now on Q will be assumed normalized in this way. A sketch of Q is shown in figure 1, the cross-hatched sections denoting entries which are the identity of H. Denote by $Q^{(\gamma)}$ the $M_\gamma \times I_\gamma$-submatrix of Q.

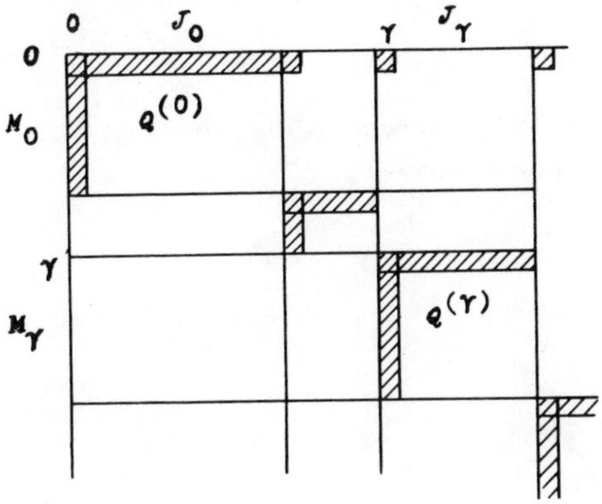

Figure 1

Using (1) it is now fairly routine to show

LEMMA 2.2. *For each* $\gamma \in \Gamma$, *let* $H_\gamma = \{h \in H\colon (h;\gamma',\gamma') \in V_\gamma\}$. *Then* H_γ *is a subgroup of* H *isomorphic with* G_γ *and* $V_\gamma = H_\gamma \times J_\gamma \times M_\gamma$. *The structure group* H *of* T *is generated by* $\cup\{H_\gamma\colon \gamma \in \Gamma\}$ *together with the entries of* Q. *In fact* H *is generated by* $\{H_\gamma\colon \gamma \in \Gamma\}$ *together with the set* $\overline{Q}$, *where*

$$\overline{Q} = \{q_{\mu j} \in Q\colon \textit{(i)}\ (\mu,j) \notin \cup\{M_\gamma \times J_\gamma\colon \gamma \in \Gamma\}\ \textit{and}$$

$$\textit{(ii)}\ (\mu,j) \neq (o',\gamma')\ \textit{for all}\ \gamma \in \Gamma\}.$$

We now show that the elements of $\overline{Q}$ may be chosen "arbitrarily" and that H is generated "in the freest possible way" by the subgroups H_γ, $\gamma \in \Gamma$, and by $\overline{Q}$. More precisely we construct the following Rees matrix semigroup S which will prove to be isomorphic with T.

Let I and Λ be the disjoint unions of the sets I_γ, $\gamma \in \Gamma$, and Λ_γ, $\gamma \in \Gamma$, respectively. (Clearly $|I| = |J|$ and $|\Lambda| = |M|$.) If $0 \in \Gamma$ is chosen as above, let

$$A(0,\Gamma) = \{(\lambda,i) \in \Lambda \times I: \ \text{(i)} \ (\lambda,i) \notin \dot{\cup}\{\Lambda_\gamma \times I_\gamma: \gamma \in \Gamma\} \ \text{and}$$
$$\text{(ii)} \ (\lambda,i) \neq (0,\gamma) \ \text{for all} \ \gamma \in \Gamma\}.$$

(Note that we may consider $\overline{Q}$ to be indexed by $A(0,\Gamma)$.) Let $\overline{P}$ be a set of distinct elements indexed by $A(0,\Gamma)$:

$$\overline{P} = \{p_{\lambda i}: (\lambda,i) \in A(0,\Gamma)\}.$$

Let G be the ($\mathcal{G}r$-) free product of the groups G_γ, $\gamma \in \Gamma$, and of the ($\mathcal{G}r$-) free group on $\overline{P}$; that is

$$G = \underset{\mathcal{G}r}{*}\{G_\gamma: \gamma \in \Gamma\} \underset{\mathcal{G}r}{*} F_{\overline{P}}(\mathcal{G}r).$$

We may treat $\overline{P}$ as a subset of G. Let P be the $\Lambda \times I$ matrix over G whose (λ,i)th entry is

$$\begin{cases} p_{\lambda i}^{(\gamma)} & \text{if } (\lambda,i) \in \Lambda_\gamma \times I_\gamma \quad \text{for some } \gamma \in \Gamma, \\ e & \text{if } (\lambda,i) = (0,\gamma) \quad \text{for some } \gamma \in \Gamma, \\ p_{\lambda i} & \text{otherwise.} \end{cases} \tag{2}$$

(Note that P has the same form as the matrix in figure 1 .)

THEOREM 2.3. *The semigroup $S = M(G;I,\Lambda;P)$ is isomorphic with the $\mathcal{CS}$-free product of the family $U_\gamma = M(G_\gamma;I_\gamma,\Lambda_\gamma;P^{(\gamma)})$, $\gamma \in \Gamma$.*

To complete the proof of this theorem observe first that we may treat each U_γ as the (completely simple) subsemigroup $U_\gamma = G_\gamma \times I_\gamma \times \Lambda_\gamma$ of S under the multiplication determined by P. Then

LEMMA 2.4. *The semigroup S is generated, as a completely simple semigroup, by $\cup\{U_\gamma: \gamma \in \Gamma\}$.*

By universality of $T = \underset{\mathcal{CS}}{*}\{U_\gamma: \gamma \in \Gamma\}$, then, there is a morphism θ of T upon S such that $\iota_\gamma\theta$ is the identity on U_γ for each $\gamma \in \Gamma$. (In other words θ extends the morphisms ι_γ^{-1} of V_γ into S.)

On the other hand from the universality of G it follows from lemma 2.2 that there is a morphism ω of G upon H extending the morphisms $\omega_\gamma: G_\gamma \to H_\gamma$ induced by ι_γ, $\gamma \in \Gamma$, and mapping $p_{\lambda i}$ to $q_{\lambda' i'}$ for each $p_{\lambda i} \in \overline{P}$ (that is $(\lambda,i) \in A(0,\Gamma)$). It is now routine to verify that ω extends to a morphism of $S = M(G;I,\Lambda;P)$ upon $T = M(H;J,M;Q)$ extending the morphism ι_γ, $\gamma \in \Gamma$. Hence ψ and θ are mutually inverse. This completes the proof of theorem 2.3.

In sections 4 and 5 we will give two applications of theorem 2.3.

3. Amalgamation

Let $\{U_\gamma : \gamma \in \Gamma\}$ be a family of completely simple semigroups with a common completely simple subsemigroup V. Clarke [A] has modified a construction due to Imaoka [1977] to show that every such amalgam $[U_\gamma : \gamma \in \Gamma; V]$ is *strongly embeddable* within $\underset{\sim}{CS}$, that is there is a semigroup U in $\underset{\sim}{CS}$ and *embeddings* $\pi_\gamma : U_\gamma \to U$, $\gamma \in \Gamma$, agreeing on V, such that $U_\gamma\pi_\gamma \cap U_\varepsilon\pi_\varepsilon = V\pi_\gamma (= V\pi_\varepsilon)$ for all distinct $\gamma, \varepsilon \in \Gamma$. Thus $\underset{\sim}{CS}$ has the *strong amalgamation property*.

Let T be the *$\underset{\sim}{CS}$-free product of the amalgam* $[U_\gamma : \gamma \in \Gamma; V]$, defined in terms of the usual universal properties. (Since CS is a variety, such a T exists). Then there exist morphisms $\eta_\gamma : U_\gamma \to T$, $\gamma \in \Gamma$, agreeing on V, such that for any family of morphisms $\theta_\gamma : U_\gamma \to W$, $\gamma \in \Gamma$, agreeing on V, ($W \in \underset{\sim}{CS}$), there is a morphism $\theta : T \to W$ such that $\eta_\gamma\theta = \theta_\gamma$, $\gamma \in \Gamma$. In view of the previous paragraph (see, for instance [Howie 1976, VII.1]) the morphisms η_γ are embeddings such that $U_\gamma\eta_\gamma \cap U_\varepsilon\eta_\varepsilon = V\eta_\varepsilon$ for all distinct $\gamma, \varepsilon \in \Gamma$.

Following Imaoka [1977a] we may choose $U_\gamma = M(G_\gamma; I_\gamma, \Lambda_\gamma; P^{(\gamma)})$, $\gamma \in \Gamma$, and $V = M(L;K,\Delta;R)$ in such a way that for all distinct $\gamma, \varepsilon \in \Gamma$, $I_\gamma \cap I_\varepsilon = K$, $\Lambda_\gamma \cap \Lambda_\varepsilon = \Delta$, $G_\gamma \cap G_\varepsilon = L$ and $p^{(\gamma)}_{\delta k} = r_{\delta k}$ for all $(\delta, k) \in \Delta \times K$. We may also suppose that each $P^{(\gamma)}$ is normalized with respect to some $1 \in K$, $1 \in \Delta$.

Let $I = \cup\{I_\gamma : \gamma \in \Gamma\}$ and $\Lambda = \cup\{\Lambda_\gamma : \gamma \in \Gamma\}$. Let $\hat{P}$ be a set indexed by $\cup\{\Lambda_\varepsilon \times I_\gamma : \gamma, \varepsilon \in \Gamma,\ \gamma \neq \varepsilon\}$, say

$$\hat{P} = \{p_{\lambda i} : (\lambda, i) \in \cup\{\Lambda_\varepsilon \times I_\gamma : \gamma \neq \varepsilon\}\}.$$

Let $G = A \underset{\underset{\sim}{Gp}}{*} F_{\hat{P}}(\underset{\sim}{Gp})$, where A is the $\underset{\sim}{Gp}$-free product of the amalgam $[G_\gamma : \gamma \in \Gamma; L]$. (We may treat $\hat{P}$ and the groups G_γ as subsets of G.) Let P be the $\Lambda \times I$ matrix over G whose (λ, i) entry is

$$\begin{cases} p_{\lambda i}^{(\gamma)} & \text{if } (\lambda, i) \in \Lambda_\gamma \times I_\gamma,\ \gamma \in \Gamma \\ p_{\lambda i} & \text{otherwise.} \end{cases}$$

A sketch of the matrix P is shown in figure 2, the shaded regions denoting the entries in $\hat{P}$, and the cross-hatched regions entries being the identity.

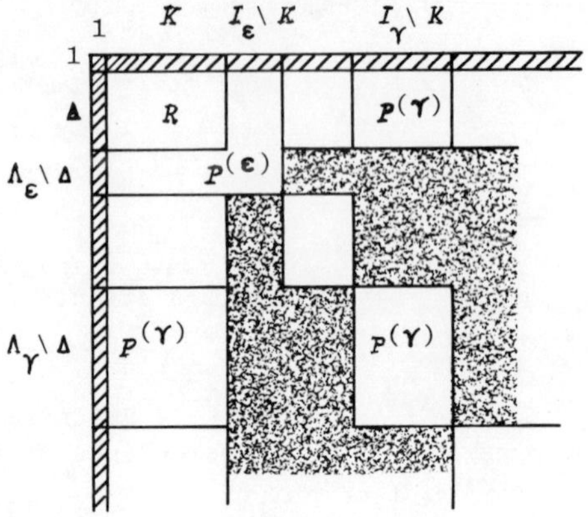

Figure 2

THEOREM 3.1. *In the above notation,* $S = M(G;I,\Lambda;P)$ *is the* $\mathbf{CS}$*-free product of the amalgam* $[U_\gamma : \gamma \in \Gamma; V]$.

The proof of the theorem follows closely that of theorem 2.3 and we omit the details.

Varieties of completely simple semigroups with the strong amalgamation property are considered by Clarke [A] elsewhere in this volume.

4. Free completely simple semigroups

If X is a non-empty set and $\mathbf{V}$ is a variety of completely simple semigroups denote by $F_X(\mathbf{V})$ the $\mathbf{V}$-free semigroups over X. It is immediate from universality that $F_X(\mathbf{V})$ is isomorphic with the $\mathbf{V}$-free product of a family $\{U_x : x \in X\}$ of isomorphic copies of $F_1(\mathbf{V})$. In particular $F_X(\mathbf{CS})$ is isomorphic with $\underset{\mathbf{CS}}{*}\{U_x : x \in X\}$, where each U_x is isomorphic with the $\mathbf{Gp}$-free cyclic group. Writing $U_x = M(G^{(x)};\{x\},\{x\};(e))$ we may use theorem 2.3 to give a Rees matrix description for $F_X(\mathbf{CS})$.

After normalizing the resulting matrix with respect to the row and column corresponding to $0 \in X$ (in the terminology of the theorem) we obtain the following description, first obtained by Clifford [1979a].

THEOREM 4.1. *Let X be a non-empty set. Pick some element 0 of X, let $\overline{Q} = \{q_{yx}: (y,x) \in (X\setminus 0) \times (X\setminus 0)\}$ be a set of cardinality $|X\setminus 0|^2$ and let Q be the $X \times X$ matrix whose entry in (y,x)th place is*

$$\begin{cases} e & \text{if } y = 0 \text{ or } x = 0 \\ q_{yx} & \text{otherwise.} \end{cases}$$

Let H be the ($\mathcal{G}p$-) free group on $X \cup \overline{Q}$. Then

$$F_X(\mathcal{CS}) \simeq M(H;X,X;Q).$$

It follows from the theorem that if we treat, as we may, the set X as actually a subset of H, then $F_X(\mathcal{CS})$ is freely generated by $\{(x;x,x): x \in X\}$ in the above representation. Note also that the maximal subgroups of the $\mathcal{CS}$-free semigroup on n generators (n finite) are $\mathcal{G}p$-free on $n + (n - 1)^2$ generators.

Now let α, β be non-zero cardinal numbers. Eberhart, Kinch and Williams [1973a] showed that there exists an idempotent-generated completely simple semigroup $T(\alpha,\beta)$ with β L-classes and α R-classes and having the property that *every* idempotent-generated completely simple semigroup S with β L-classes and α R-classes is a morphic image of $T(\alpha,\beta)$. In fact if κ and τ are bijections of $T(\alpha,\beta)/L$ upon S/L and of $T(\alpha,\beta)/R$ upon S/R respectively, there is a morphism θ of $T(\alpha,\beta)$ upon S with the property that for every L-class L_λ and R-class R_i of $T(\alpha,\beta)$, $L_\lambda\tau = L_{\lambda\kappa}$ and $R_i\theta = R_{i\tau}$.

It is easily verified that $T(\alpha,\beta)$ is the unique semigroup, up to isomorphism, with these properties. We call $T(\alpha,\beta)$ the *free idempotent-generated completely simple semigroup with β L-classes and α R-classes.* (In the terminology of [Pastijn, H] and [Meakin, F], $T(\alpha,\beta)$ is the "free regular idempotent-generated semigroup over the $\beta \times \alpha$ rectangle" (considered as a biordered set).)

It was conjectured by Eberhart, Kinch and Williams that the maximal subgroups of $T(\alpha,\beta)$ are always free and this was essentially proved in [Pastijn 1977b]. It is an immediate consequence of

THEOREM 4.2. *Let α and β be non-zero cardinals with $\alpha \geqslant \beta$. Let $Y \subseteq X$ be sets of cardinality β, α respectively. Then $T(\alpha,\beta)$ is isomorphic with the subsemigroup of $F_X(\underset{\sim}{CS})$ generated by the idempotents e_{xy}, $(x,y) \in X \times Y$ (in the notation of theorem 4.1).*

Proof. Let S be an idempotent-generated completely simple semigroup with β L-classes and α R-classes. We may suppose $S = M(K;X,Y;R)$, where R is normalized with respect to $0 \in Y(\subseteq X)$. By [Howie 1978], $K = \langle R \rangle$, the subgroup generated by the entries of R. If f_{xy} denotes the idempotent of S in H_{xy}, then a proof similar to that used in lemma 2.4 shows that S is generated by

$$\{f_{xx}\colon x \in Y\} \cup \{f_{x0}\colon x \in X\setminus Y\}, \text{ with cardinality } |X|.$$

By theorem 4.1 there is a morphism θ of $F_X(\underset{\sim}{CS})$ upon S such that

$$(x;x,x)\theta = \begin{cases} f_{xx} & \text{if } x \in Y \\ f_{x0} & \text{if } x \in X\setminus Y. \end{cases}$$

It then follows that $e_{xy}\theta = f_{xy}$ for all $(x,y) \in X \times Y$, so that θ maps the subsemigroup of $F_X(\underset{\sim}{CS})$ generated by $\{e_{xy}\colon (x,y) \in X \times Y\}$ onto S. It is then not difficult to verify that this subsemigroup satisfies the remaining requirements.

COROLLARY 4.3. *$T(\alpha,\beta) \simeq M(\langle P_{Y,X}\rangle;X,Y;P_{Y,X})$, where $P_{Y,X}$ is the $Y \times X$-submatrix of P (and we assume P is normalized with respect to $0 \in Y$). Thus the maximal subgroups of $T(\alpha,\beta)$ are free on a set of cardinality $|X\setminus 0|.|Y\setminus 0|$.*

5. Injectives and projectives in $\underset{\sim}{CS}$

An object A in a category C is a C-*injective* if for every monomorphism $\alpha\colon B \to C$ and every morphism $\beta\colon B \to A$ in C, there exists a morphism $\gamma\colon C \to A$ in C so that $\alpha\gamma = \beta$.

If S is a $\underset{\sim}{CS}$-injective it is easy to see that each maximal subgroup of S is a $\underset{\sim}{Gp}$-injective, whence trivial (see, for instance, [Schein 1974]). Thus S is a rectangular band. Conversely any rectangular band is a $\underset{\sim}{CS}$-injective.

PROPOSITION 5.1. *The* $\mathcal{CS}$*-injectives are precisely the rectangular bands.*

The *C-projectives* are defined dually: A is a C-projective if for every epimorphism $\alpha: C \to B$ and every morphism $\beta: A \to B$ in C, there exists a morphism $\gamma: A \to C$ so that $\gamma\alpha = \beta$. If C is a *variety of algebras* then the C-projectives are precisely the *retracts* of C-free algebras. (We shall call a subalgebra P of an algebra A a retract of A if there is a morphism of A into P whose restriction to P is the identity.) Further, the C-free product of C-projectives is again a C-projective. Note that since $\mathcal{CS}$ has the strong amalgamation property (see §3), the epimorphisms in $\mathcal{CS}$ are just the surjective morphisms.

THEOREM 5.2. *A semigroup* S *is a* $\mathcal{CS}$*-projective if and only if* $S \simeq M(K;J,M;R)$*, where* R *is normalized with respect to some row and column and* K *is a retract of some group* H*, freely generated by a set containing the remaining entries of* R*.*

Proof. If S is a $\mathcal{CS}$-projective then S is a retract of a $\mathcal{CS}$-free semigroup F on some set X. From theorem 4.1, we may choose $F = M(H;X,X;Q)$, where Q is normalized with respect to some $0 \in X$ and H is freely generated by X and the remaining entries of Q.

Observe that (by renormalizing, if necessary) $S = M(K;J,M;R)$, where J and M are subsets of X containing 0, R is the $M \times J$-submatrix of Q and $K = \{h \in H: (h,0,0) \in S\}$ is a subgroup of H. In fact the retraction of F upon S induces a retraction of H upon K and H is clearly freely generated by a set containing those entries of R outside the normalized row and column.

Conversely, note first that any left or right group whose maximal subgroups are either free or trivial is a $\mathcal{CS}$-projective. Now let $S = M(K;J,M;R)$ be as in the statement of the theorem. The retraction of H upon K extends to a retraction of $T = M(H;J,M;R)$ upon S. If $|J| = 1$ or $|M| = 1$ then T is $\mathcal{CS}$-projective so that S is also. Suppose that $|J| > 1$, $|M| > 1$, and that R is normalized with respect to $0 \in J$, $0 \in M$. Let $\overline{R}$ be the set of entries of R outside that column and row. From theorem 2.3 we see that T is isomorphic to the $\mathcal{CS}$-free product of the left zero semigroup on $J\setminus 0$ and the right group $(M\setminus 0) \times F_{Y\setminus \overline{R}}(\mathcal{Gp})$, where Y is the given set of free generators of H. Since the $\mathcal{CS}$-product preserves projectivity, T is projective, whence S is also.

COROLLARY 5.3. *The orthodox $\underset{\sim}{CS}$-projectives are the left and right groups with trivial or free subgroups.*

The non-orthodox $\underset{\sim}{CS}$-projectives are the full retracts of $\underset{\sim}{CS}$-free products of left zero semigroups and right groups (with trivial or free subgroups).

Clearly any retract of a free group is again free — the precise nature of the retracts is given in [Magnus, Karrass and Solitar 1966].

6. The lattice of varieties

In any variety $\underset{\sim}{V}$ of algebras the lattice $V(\underset{\sim}{V})$ of subvarieties is dually isomorphic with the lattice of fully invariant congruences on the free algebra on a countably infinite set of generators in that variety, and moreover the lattice of fully invariant congruences on any algebra A forms a sublattice of the lattice of congruences on A. (See, for example [B.H. Neumann 1961]. A congruence ρ on an algebra A is *fully invariant* if $a\rho b$ implies $a\theta\rho b\theta$ for every endomorphism θ of A.)

From this point of view it is not difficult to show

THEOREM 6.1. *The lattice $V(\underset{\sim}{CS})$ of varieties of completely simple semigroups is modular.*

Proof. From theorem 3.9 of [Spitznagel 1973b] it follows that the lattice $\Lambda(S)$ of congruences on any completely simple semigroup S is isomorphic with a subdirect product of the lattices $[\iota, H]$ (consisting of those congruences contained in H) and $\Lambda(S/H)$, where S/H is a rectangular band. In particular, this decomposition respects full invariance — thus the lattice of fully invariant congruences on S is a subdirect product of a sublattice of $[\iota, H]$ (modular, by [Munn 1964]) with the lattice of fully invariant congruences on a rectangular band. The latter is easily seen to be distributive, whence the lattice of fully invariant congruences on S is modular. Since modularity is a self-dual concept, $V(\underset{\sim}{CS})$ is modular.

In fact, more generally T.E. Hall and the author [B] have shown, using these methods, that the lattice of varieties of bands of groups (completely regular semigroups on which H is a congruence) is modular.

Now we consider the lattice $V(\mathbf{CS})$ in more detail. We first decompose $V(\mathbf{CS})$ into the disjoint union of the interval sublattices $[\mathbf{G},\mathbf{CS}(\mathbf{G})]$, $\mathbf{G} \in V(\mathbf{Gp})$, where $\mathbf{CS}(\mathbf{G}) = \{S \in \mathbf{CS}$: the subgroups of S belong to $\mathbf{G}\}$. Clearly $[\mathbf{G},\mathbf{CS}(\mathbf{G})] = \{\mathbf{V} \in V(\mathbf{CS})\colon \mathbf{V} \cap \mathbf{Gp} = \mathbf{G}\}$. Covering $\mathbf{G}$ in this interval are the varieties $\mathbf{Left}(\mathbf{G})$ of *left groups over* $\mathbf{G}$ and $\mathbf{Right}(\mathbf{G})$ of *right groups over* $\mathbf{G}$, whose join is $\mathbf{RG}(\mathbf{G})$, the variety of *rectangular groups over* $\mathbf{G}$. It is easily verified that $\mathbf{RG}(\mathbf{G}) = \mathbf{G} \vee \mathbf{RB}$ and that $\mathbf{Left}(\mathbf{G}) = \mathbf{G} \vee \mathbf{LZ}$ and $\mathbf{Right}(\mathbf{G}) = \mathbf{G} \vee \mathbf{RZ}$ (where $\mathbf{RB}$, $\mathbf{LZ}$ and $\mathbf{RZ}$ denote the varieties of rectangular bands, left zero and right zero semigroups respectively).

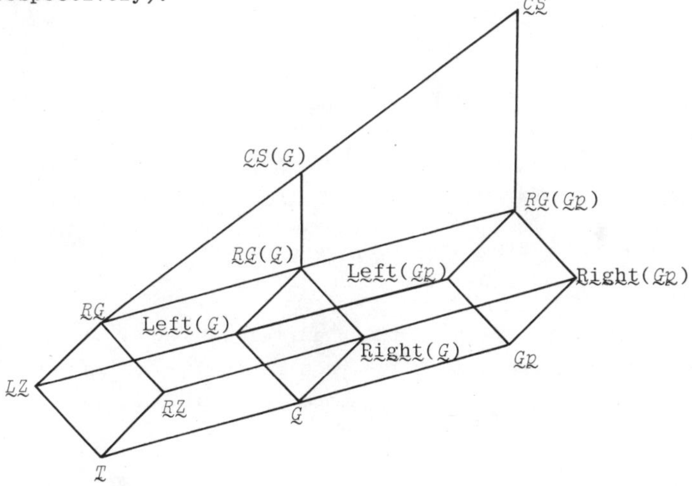

Figure 3

From now on we consider the interval $[\mathbf{RG}(\mathbf{G}),\mathbf{CS}(\mathbf{G})]$, $\mathbf{G} \in V(\mathbf{Gp})$. Let F denote the $\mathbf{CS}(\mathbf{G})$-free semigroup on countably infinitely many generators. The subvariety $\mathbf{G}$ corresponds to the fully invariant congruence σ (the least group congruence) on F; $\mathbf{RB}$ corresponds to the fully invariant congruence H. Hence $\mathbf{RG}(\mathbf{G})$ corresponds to $H \cap \sigma$. Thus the interval $[\mathbf{RG}(\mathbf{G}),\mathbf{CS}(\mathbf{G})]$ is dually isomorphic with the lattice of fully invariant congruences on F *contained in* $H \cap \sigma$. We now describe these congruences in terms of certain subgroups of the structure group of F.

Suppose F is free on the set A. As noted in the introduction theorem 4.1 has a direct analogue: we may choose

$$F = M(H;I,\Lambda;Q),$$

where $|I| = |\Lambda| = |A|$, and where H is freely generated *in* $\mathbf{G}$ by A together with the set $\overline{Q}$ of non-trivial entries of Q, the matrix being normalized with respect to some 0th row and 0th column. Denoting by G_X the $\mathcal{G}$-free group on a set X, we have $H = G_{A\cup\overline{Q}} = G_A \underset{\mathcal{G}}{*} G_{\overline{Q}}$.

The congruences on F contained in $\mathcal{H}$ can be described as follows (see, for example, [Howie 1976, §III.4]). For each congruence ρ on F let $N_\rho = \{a \in H: (a;0,0)\ \rho\ (e;0,0)\}$. Then N_ρ is a normal subgroup of H and the map $\rho \to N_\rho$, $\rho \subseteq \mathcal{H}$, is an isomorphism of the lattice of congruences on F contained in $\mathcal{H}$ upon the lattice of normal subgroups of H. Further,

$$F/\rho \simeq \mathcal{M}(H/N_\rho;I,\Lambda;Q/N_\rho),$$

where Q/N_ρ is the $\Lambda \times I$ matrix over H/N_ρ whose entries are the cosets of the entries in Q. Now from Exercise 3.4.1 of [Clifford and Preston 1961], $N_\sigma = (G_{\overline{Q}})^H$, the normal closure of the subgroup generated by the entries of Q. Thus

LEMMA 6.2. *The lattice of congruences on F contained in $\mathcal{H} \cap \sigma$ is isomorphic with the lattice of normal subgroups of H contained in $(G_{\overline{Q}})^H$.*

To determine which of these normal subgroups correspond to fully invariant congruences on F we require the following description of the endomorphisms of F, based on theorem 3.11 of [Clifford and Preston 1961].

LEMMA 6.3. *An endomorphism of F is determined by an endomorphism ω of H and transformations ψ of I, χ of Λ such that for every $(\lambda,i) \in (\Lambda\setminus 0) \times (I\setminus 0)$,*

$$q_{\lambda i}\omega = u_0^{-1}(q_{\lambda\lambda,0\psi}^{-1}\, q_{\lambda\chi,i\psi}\, q_{0\chi,i\psi}^{-1}\, q_{0\chi,0\psi})u_0, \text{ for some } u_0 \in H. \qquad (3)$$

We will say an endomorphism ω of H is *Q-admissible* if there are transformations ψ of I, χ of Λ and an element u_0 of H such that (3) is satisfied for every $q_{\lambda i} \in \overline{Q}$. (The equation reflects the obvious fact that, although endomorphisms of H in general may map the generators $q_{\lambda i}$ arbitrarily, endomorphisms of F must map $\mathcal{L}$-[$\mathcal{R}$-] related idempotents to $\mathcal{L}$-[$\mathcal{R}$-] related idempotents. Note also that the bracketed term in (3) is an "extract" of Q, in the terminology of [Howie 1976]).

If N is a subgroup of H we say N is *Q-invariant* if $N\omega \subseteq N$ for every Q-admissible endomorphism ω of H. The following elementary properties of such subgroups may be routinely established.

LEMMA 6.4. *(i) Any Q-invariant subgroup of H is normal in H.*

(ii) If N is a fully invariant subgroup of $G_{\overline{Q}}$ *then* N^H *is a Q-invariant subgroup of H contained, in particular, in the Q-invariant subgroup* $(G_{\overline{Q}})^H$.

(iii) If ρ is a congruence on F contained in H, then ρ is fully invariant if and only if N_ρ *is Q-invariant.*

Observe finally that if M and N are Q-invariant subgroups of H then so are MN and $M \cap N$; that is, the lattice of Q-invariant subgroups of H is a sublattice of the lattice of normal subgroups of H. We now collect our results into a theorem which describes the lattice $V(\underset{\sim}{CS})$ in terms of the Q-invariant subgroups of the groups $G_{A\cup\overline{Q}}$, for group varieties $\underset{\sim}{G}$.

THEOREM 6.5. *The lattice* $V(\underset{\sim}{CS})$ *of varieties of completely simple semigroups is the disjoint union of the interval sublattices* $[\underset{\sim}{G}, \underset{\sim}{CS}(\underset{\sim}{G})]$, $\underset{\sim}{G} \in V(\underset{\sim}{Gp})$. *Each sublattice* $[\underset{\sim}{G}, \underset{\sim}{CS}(\underset{\sim}{G})]$ *has the form described in figure 3. Its sublattice* $[\underset{\sim}{RG}(\underset{\sim}{G}), \underset{\sim}{CS}(\underset{\sim}{G})]$ *is dually isomorphic with the lattice of fully invariant congruences, contained in* $\sigma \cap H$, *on* $F_A = M(H;I,\Lambda;Q)$, *(the* $\underset{\sim}{CS}(\underset{\sim}{G})$*-free semigroup on countably many generators), which in turn is isomorphic with the lattice of Q-invariant subgroups of* $(G_{\overline{Q}})^H$, *the normal closure of* $G_{\overline{Q}}$ *in* $H = G_{A\cup\overline{Q}}$, *via the isomorphism* $\rho \to N_\rho$.

Thus every subvariety $\underset{\sim}{V}$ of $\underset{\sim}{CS}$ containing $\underset{\sim}{RB}$ is determined by $\underset{\sim}{G} = \underset{\sim}{V} \cap \underset{\sim}{Gp}$ and a Q-invariant subgroup N, say, of $H = G_{A\cup\overline{Q}}$ contained in $(G_{\overline{Q}})^H$. If $\underset{\sim}{W} \in V(\underset{\sim}{CS})$ and $\underset{\sim}{W}$ also contains $\underset{\sim}{RB}$, let $\underset{\sim}{K} = \underset{\sim}{W} \cap \underset{\sim}{Gp}$ and let M be the associated Q-invariant subgroup of $K_{A\cup\overline{Q}}$. Then $\underset{\sim}{V} \subseteq \underset{\sim}{W}$ if and only if $\underset{\sim}{G} \subseteq \underset{\sim}{H}$ and N is contained in the image of M under the natural morphism of $K_{A\cup\overline{Q}}$ upon $G_{A\cup\overline{Q}}$.

EXAMPLE 6.6. *The varieties* $\underset{\sim}{CS}(\underset{\sim}{G},\underset{\sim}{K}), \underset{\sim}{K} \subseteq \underset{\sim}{G}$. Let N be a *fully* invariant subgroup of $G_{\overline{Q}}$. As noted in lemma 6.4, N^H is a Q-invariant subgroup of H. Let K be the subvariety of G determined by N (see [H. Neumann 1967b; p.12]) and denote by $\underset{\sim}{CS}(\underset{\sim}{G},\underset{\sim}{K})$ the subvariety of $\underset{\sim}{CS}(\underset{\sim}{G})$ corresponding to N^H. From the comments preceding lemma 6.2 we see that $F_A(\underset{\sim}{CS}(\underset{\sim}{G},\underset{\sim}{K})) \simeq M(H/N^H;I,\Lambda;Q/N^H)$, where in fact

$$H/N^H \simeq G_A \underset{\underset{\sim}{G}}{*} K_{\overline{Q}/N^H},$$

($K_{\overline{Q}/N^H}$ denoting the $\underset{\sim}{K}$-free group on $\overline{Q}/N^H$), using the fact ([H. Neumann 1967b, p.36]) that in any $\underset{\sim}{G}$-free product the normal closure of one free factor avoids the other. It can be shown that

$$\underset{\sim}{CS}(\underset{\sim}{G},\underset{\sim}{K}) = \{S \in \underset{\sim}{CS}(\underset{\sim}{G}): \langle E_S \rangle \in \underset{\sim}{CS}(\underset{\sim}{K})\}.$$

Examples are given in [Jones, C] to show that in general a Q-invariant subgroup N of H need not be fully invariant: in fact neither need $N \cap G_{\overline{Q}}$ be a fully invariant subgroup of $G_{\overline{Q}}$. Thus not every variety in $[\underset{\sim}{RG}(\underset{\sim}{G}),\underset{\sim}{CS}(\underset{\sim}{G})]$ need in general have the form $\underset{\sim}{CS}(\underset{\sim}{G},\underset{\sim}{K})$ for some subvariety $\underset{\sim}{K}$ of $\underset{\sim}{G}$.

However if $\underset{\sim}{G}$ is a variety of *abelian* groups, then $(H=)G_{A\cup\overline{Q}} = G_A \times G_{\overline{Q}}$, now, so every Q-invariant subgroup of H is a subgroup of $G_{\overline{Q}}$ itself. By suitable choices of mappings ω, ψ and χ in equation (3) it can be shown that the Q-invariant subgroups of H contained in $G_{\overline{Q}}$ are *precisely* the fully invariant subgroups of $G_{\overline{Q}}$. Hence we obtain

THEOREM 6.7 ([Rasin 1979b]). *A variety $\underset{\sim}{V}$ of completely simple semigroups with abelian subgroups is either*

(i) a variety of abelian groups,

or (ii) a variety of left or right abelian groups,

or (iii) of the form $\underset{\sim}{CS}(\underset{\sim}{G},\underset{\sim}{H})$ for some pair $\underset{\sim}{H} \subseteq \underset{\sim}{G}$ of varieties of abelian groups, where $\underset{\sim}{CS}(\underset{\sim}{G},\underset{\sim}{H}) = \{S \in \underset{\sim}{CS}(\underset{\sim}{G}): \langle E_S \rangle \in \underset{\sim}{CS}(\underset{\sim}{H})\}$.

Since every variety of abelian groups is either $\underset{\sim}{Ab}$ (the variety of abelian groups) or $\underset{\sim}{A}_k$ (defined within $\underset{\sim}{Ab}$ by $x^k = 1$), $k \geq 1$, this theorem describes the varieties completely, for defining identities may be easily deduced. For instance if $k|m$, $\underset{\sim}{CS}(\underset{\sim}{A}_m,\underset{\sim}{A}_k)$ is defined within $\underset{\sim}{CS}(\underset{\sim}{Ab})$ by

$$(xx^{-1}yy^{-1})^k = (xy)(xy)^{-1} \text{ and } x^m = xx^{-1}.$$

Although theorem 6.5 is clearly useful in the above setting, in general the group-theoretic complications which quickly arise make the prospect of a similarly neat description of $V(\underset{\sim}{CS})$ itself very discouraging.

Notice finally that implicit in theorem 6.5 is a description (using the earlier comments) of the $\underset{\sim}{V}$-free semigroups on the (countable) set A, where $\underset{\sim}{V} \in [\underset{\sim}{RG}(\underset{\sim}{G}), \underset{\sim}{CS}(\underset{\sim}{G})]$. A similar description may be given of the $\underset{\sim}{V}$-free semigroups on an *arbitrary* set X in terms of the $\underset{\sim}{CS}(\underset{\sim}{G})$-free semigroup on X. (For varieties $\underset{\sim}{V}$ of left [right] groups, $F_X(\underset{\sim}{V})$ is the direct product of $F_X(\underset{\sim}{V} \cap \underset{\sim}{Gp})$ with the left zero [right zero], semigroup over X.)

Similarly a description of the *$\underset{\sim}{V}$-free idempotent-generated completely simple semigroup with* β *L-classes and* α *R-classes* may be given, as a subsemigroup of a certain $\underset{\sim}{V}$-free semigroup (cf section 4). In general the maximal subgroups of such a semigroup need *not* be free in any variety of groups (see [Jones, C, §6]).

7. The Hopf property

An algebra is called *Hopfian* if each of its onto endomorphisms is an automorphism. H. Neumann [1967b] has posed the following problem:

Is $F_X(\underset{\sim}{G})$ Hopfian for every variety $\underset{\sim}{G}$ of groups and every finite set X?

We may of course ask the analogous question for all varieties of completely simple semigroups, or for that matter, for varieties of algebras in general. T. Evans [1971] posed the analogous question for varieties of *semigroups*. We have shown [Jones, D] that the answer is yes for any variety of semigroups not consisting entirely of completely simple semigroups; thus the only remaining cases to consider in Evans' problem are subvarieties of $\underset{\sim}{CS}(\underset{\sim}{B}_k)$, $k \geqslant 1$, where $\underset{\sim}{B}_k$ denotes the Burnside variety of groups (defined by $x^k = 1$).

Although, as mentioned in the previous section, we have a Rees matrix representation for $F_X(\underset{\sim}{V})$, where $\underset{\sim}{V} \in V(\underset{\sim}{CS})$, we have been unable to use this description to answer the above question for all such varieties $\underset{\sim}{V}$. However we do have the following partial result.

THEOREM 7.1. *Let $\underset{\sim}{G}$ be a variety of groups. If every $\underset{\sim}{G}$-free group of finite rank is Hopfian then so is every $\underset{\sim}{CS}(\underset{\sim}{G})$-free semigroup of finite rank. In particular every $\underset{\sim}{CS}$-free semigroup of finite rank is Hopfian.*

Proof. According to the analogue of theorem 4.1, the maximal subgroups of $F_n(\underset{\sim}{CS}(\underset{\sim}{G}))$ are isomorphic to the $\underset{\sim}{G}$-free groups of rank

$n + (n - 1)^2$. By corollary 7.3 of [Jones (E)] a finitely generated completely simple semigroup whose maximal subgroups are Hopfian is itself Hopfian and the first statement now follows. In particular it is known that all (absolutely) free groups of finite rank are Hopfian, from which the second statement follows.

We remark that the converse of the result quoted in the proof is not true: there exist Hopfian completely simple semigroups whose maximal subgroups are non-Hopfian (see [Jones (E), §7]).

Ideally, the answer to the analogue of Neumann's problem for subvarieties of $\underset{\sim}{CS}$ would have the following answer.

CONJECTURE. *If* $\underset{\sim}{V}$ *is a variety of completely simple semigroups then* $F_n(\underset{\sim}{V})$ *is Hopfian if and only if the* $\underset{\sim}{G}$*-free group of rank* n*, where* $\underset{\sim}{G} = \underset{\sim}{V} \cap \underset{\sim}{Gp}$*, is Hopfian.*

References

1940 D. Rees, On semigroups. *Proc. Cambridge Phil. Soc.*, 36 (1940), 387-400.

1946 P. Whitman, Lattices, equivalence relations and subgroups. *Bull. Amer. Math. Soc.*, 52 (1946), 507-522.

1961 A.H. Clifford and G.B. Preston, *The Algebraic Theory of Semigroups*. Vol.I, Amer. Math. Soc. Surveys, Providence, R.I., 1961.

1961 B.H. Neumann, *Universal Algebra*. Lecture notes, Courant Inst. of Math. Sci., New York University, 1961.

1964 W.D. Munn, A certain sublattice of the lattice of congruences on a regular semigroup. *Proc. Cambridge Phil. Soc.*, 60 (1964), 385-391.

1966 W. Magnus, A. Karrass and D. Solitar, *Combinatorial Group Theory*. Interscience Publishers, N.Y., 1966.

1967a G. Lallement, Demi-groupes réguliers. Ann. Mat. pura ed appl., 77 (1967), 47-129.

1967b H. Neumann, *Varieties of Groups*. Springer Verlag, N.Y., 1967.

1968 D.B. McAlister, A homomorphism theorem for semigroups. *J. London Math. Soc.*, 43 (1968), 355-366.

1971 T. Evans, The lattice of semigroup varieties. *Semigroup Forum*, 2 (1971), 1-43.

1973a C. Eberhart, W. Williams and L. Kinch, Idempotent-generated regular semigroups. *J. Austral. Math. Soc.*, 15 (1973), 27-34.

1973b C. Spitznagel, The lattice of congruences on a band of groups. *Glasgow Math. J.*, 14 (1973), 189-197.

1974 B.M. Schein, Injectives in certain classes of semigroups. *Semigroup Forum*, 9 (1974), 159-171.

1975 M. Petrich, Varieties of orthodox bands of groups. *Pacific J. Math.*, 58 (1975), 209-217.

1976 J.M. Howie, *An Introduction to Semigroup Theory*. Academic Press, London, 1976.

1977a T. Imaoka, Free products with amalgamation of semigroups.

1977b F. Pastijn, Idempotent-generated completely 0-simple semigroups. *Semigroup Forum*, 15 (1977), 41-50.

1977c M. Petrich, Certain varieties and quasivarieties of completely regular semigroups. *Can. J. Math.*, 29 (1977), 1171-1197.

1978 J.M. Howie, Idempotents in completely 0-simple semigroups. *Glasgow Math. J.*, 19 (1978), 109-113.

1979a A.H. Clifford, The free completely regular semigroup on a set. *J. Algebra*, 59 (1979), 434-451.

1979b V.V. Rasin, On the lattice of varieties of completely simple semigroups. *Semigroup Forum*, 17 (1979), 113-122.

TO BE PUBLISHED

[A] G.T. Clarke, On completely regular semigroup varieties and the amalgamation property. "These proceedings".

[B] T.E. Hall and P.R. Jones, On the lattice of varieties of bands of groups. *Pacific J. Math.* (to appear).

[C] P.R. Jones, Completely simple semigroups: free products, free semigroups and varieties. Submitted.

[D] P.R. Jones, On a problem of Evans. Submitted.

[E] P.R. Jones, The Hopf property and $\underset{\sim}{K}$-free products of semigroups. *Semigroup Forum* (to appear).

[F] J.C. Meakin, Constructing biordered sets. "These proceedings".

[G] T.E. Nordahl and H.E. Scheiblich, Projective bands. *Algebra Universalis* (to appear).

[H] F. Pastijn, The biorder on the partial groupoid of idempotents of a semigroup. To appear.

[I] P.G. Trotter, Projectives in inverse semigroups. Submitted.

[J] P.G. Trotter, Projectives in some categories of semigroups. "These proceedings".

NOTE ADDED IN PROOF

Another independent construction of the $\underset{\sim}{CS}(\underset{\sim}{G})$-free semigroups, together with a discussion of their fully invariant congruences (c.f.§6) has recently been given by V.V. Rasin ('Free completely simple semigroups', *Research in contemporary algebra, Matem Zapiski (Sverdlovsk)* 1979, 140-151), in a sequel to [1979 b]. The author thanks M. Petrich for sending him a copy of this paper. Petrich and N.R.Reilly have also obtained, again independently, much of the material in §6 in 'Fully invariant congruences on the free completely simple semigroup' (manuscript).

ON THE STRUCTURE OF REGULAR SEMIGROUPS IN WHICH THE MAXIMAL SUBGROUPS FORM A BAND OF GROUPS

M. Yamada

Department of Mathematics, Shimane University,
Matsue, Shimane, Japan

Abstract

A regular semigroup S is said to be quasi-orthodox if and only if there exist an inverse semigroup I and a surjective homomorphism $f: S \to I$ such that ef^{-1} is a completely simple subsemigroup of S for each idempotent e of I. If a regular semigroup S satisfies the following property P, then S is necessarily quasi-orthodox: (P) The maximal subgroups of S form a band of groups. Such a semigroup S is called a quasi-orthodox semigroup with (P). In this paper, the structure of quasi-orthodox semigroups with (P) is studied. Structure theorems are established for the class of general quasi-orthodox semigroups and for some special classes of quasi-orthodox semigroups. In particular the concept of spined product of orthodox semigroups with (P) is introduced, and it is shown that an orthodox semigroup S is isomorphic to the spined product of an $\mathcal{H}$-degenerated orthodox semigroup and an $\mathcal{H}$-compatible inverse semigroup if and only if S has the property (P).

As generalizations of groups, there are two important basic classes of regular semigroups. One is the class of inverse semigroups, and the other is the class of completely simple semigroups. These two classes are generalized to the class of orthodox semigroups and that of completely regular semigroups respectively, and quite a lot of papers concerning these two classes have appeared during the last two decades (for example, see [Hall 1969, 1970a, 1971] and [the author 1967c, 1970b], etc. for orthodox semigroups; and [Clifford 1941], [Petrich 1967b] and [Lallement 1967a], etc. for completely regular semigroups). As a class containing both the class of orthodox semigroups and the

ISBN 0-12-319450-4

class of completely regular semigroups, the author has introduced the concept of a quasi-orthodox semigroup in his paper [A], and discussed the construction of these semigroups. In this paper, the author mainly deals with regular semigroups S having the following property:

(P) The maximal subgroups of S form a band of groups.

These systems are necessarily quasi-orthodox semigroups, as is shown later. Any H-compatible orthodox semigroup introduced by the author [1976] is one of the regular semigroups having the property (P). A structure theorem for H-compatible orthodox semigroups will be established in the last section.

Throughout this paper, the following notations and terminology will be used. For a completely regular semigroup M, the notation $M \sim \Sigma\{M_\lambda : \lambda \in \Lambda\}$ means that M is a semilattice Λ of completely simple semigroups $\{M_\lambda : \lambda \in \Lambda\}$. If an inverse semigroup Y has Λ as the semilattice of idempotents, we shall denote it by $Y(\Lambda)$. The terminology "a regular semigroup S with (P)" means "S is a regular semigroup having the property (P)". If S is a regular semigroup, $E(S)$ denotes the set of idempotents of S.

1. Basic properties

First, we shall give some necessary definitions. A semigroup T is said to be *H-compatible* if Green's H-relation H_T on T is a congruence. If every H-class of T consists of a single element, T is said to be *H-degenerated*. In this case, T is of course H-compatible. Let $I(\Delta)$ be an inverse semigroup, where Δ is the semilattice of idempotents of I, and $A \sim \Sigma\{A_\delta : \delta \in \Delta\}$ a completely regular semigroup. Then, a regular semigroup T is called a *regular extension of $A \sim \Sigma\{A_\delta : \delta \in \Delta\}$ by $I(\Delta)$* if

(1.1) A is a subsemigroup of T,

and

(1.2) there exists a surjective homomorphism $f: T \to I(\Delta)$ such that $\delta f^{-1} = A_\delta$ for each $\delta \in \Delta$.

A regular semigroup T is called a *quasi-orthodox semigroup* if there exist an inverse semigroup $I(\Delta)$ and a surjective homomorphism $f: T \to I(\Delta)$ such that δf^{-1} is a completely simple subsemigroup of T for each $\delta \in \Delta$.

Therefore, it is obvious that a regular semigroup T is a quasi-orthodox semigroup if and only if T is a regular extension of a completely regular semigroup $A \sim \Sigma\{A_\delta : \delta \in \Delta\}$ by an inverse semigroup $I(\Delta)$. T.E. Hall has shown that a regular semigroup T is a quasi-orthodox semigroup if and only if the idempotent-generated subsemigroup $\langle E(T)\rangle$ of T is a union of groups (see the author [A]).

Now, let S be a regular semigroup with (P). Let M be the union of all maximal subgroups of S. Then, $M = \cup\{H_e : e \in E(S)\}$, where $E(S)$ denotes the set of idempotents of S and H_e is the H-class (of S) containing e, and the H-classes H_e form a band (that is, M is H-compatible). Hence, it follows from [Lallement 1967a] that S is also H-compatible. Since M is a band of the maximal subgroups $\{H_e : e \in E(S)\}$, M is of course a subsemigroup of S. Hence, S is a natural regular semigroup in the sense of [Warne 1976a]. (A regular semigroup is called *natural regular* if the union of all maximal subgroups is a subsemigroup.)

Conversely, suppose that a regular semigroup S is H-compatible and natural regular. Then, the H-relation H_S on S is a congruence. Since S is also natural regular, S/H_S is an H-degenerated orthodox semigroup. Let $\xi: S \to S/H_S$ be the natural homomorphism, and put $S/H_S = W$. Since W is an H-degenerated orthodox semigroup, it follows from [the author 1976b] that W is a regular extension of $E(W) \sim \Sigma\{E_\lambda : \lambda \in \Lambda\}$ (a semilattice Λ of rectangular bands $\{E_\lambda : \lambda \in \Lambda\}$) by an H-degenerated inverse semigroup $Y(\Lambda)$. Hence, there exists a surjective homomorphism $\eta: W \to Y(\Lambda)$ such that $\lambda\eta^{-1} = E_\lambda$ for each $\lambda \in \Lambda$. Now, consider the mapping $\xi\eta : S \to Y(\Lambda)$. This $\xi\eta$ is a surjective homomorphism, and $\lambda(\xi\eta)^{-1} = E_\lambda\xi^{-1}$ for each $\lambda \in \Lambda$. Since each $E_\lambda\xi^{-1}$ is completely simple, S is a regular extension of $M \sim \Sigma\{M_\lambda : \lambda \in \Lambda\}$, where $M_\lambda = E_\lambda\xi^{-1}$ and $M = \cup\{M_\lambda : \lambda \in \Lambda\}$, by $Y(\Lambda)$. Since $e\xi^{-1}$ is an H-class (of S) containing an idempotent for each $e \in E(W)$ and since $E(W)$ is a band, it is obvious that M is a band of the H-classes $\{e\xi^{-1} : e \in E(W)\}$, that is, M is an H-compatible completely regular semigroup. Thus, S is a regular extension of the H-compatible completely regular semigroup $M \sim \Sigma\{M_\lambda : \lambda \in \Lambda\}$ by the H-degenerated inverse semigroup $Y(\Lambda)$.

Next, suppose that S is a regular extension of an H-compatible completely regular semigroup $M \sim \Sigma\{M_\lambda : \lambda \in \Lambda\}$ by an H-degenerated inverse semigroup $Y(\Lambda)$. Then, there exists a surjective homomorphism

$f: S \to Y(\Lambda)$ such that $\lambda f = M_\lambda$ for each $\lambda \in \Lambda$. Let G be a maximal subgroup of S. Since Gf is a homomorphic image of the group G, Gf must be a subgroup of $Y(\Lambda)$. Since $Y(\Lambda)$ is H-degenerated, Gf consists of a single element, that is, Gf is an idempotent of $Y(\Lambda)$. Therefore, $G \subseteq M$. This implies that M is the union of all maximal subgroups. Since M is H-compatible by assumption, M is a band of the maximal subgroups of S. That is, the maximal subgroups of S form a band of groups.

Thus, we have the following result.

THEOREM 1. *For a regular semigroup S, the following conditions are equivalent.*

(1) The maximal subgroups of S form a band of groups.

(2) S is an H-compatible natural regular (hence, H-compatible quasi-orthodox) semigroup.

(3) S is a regular extension of an H-compatible completely regular semigroup by an H-degenerated inverse semigroup.

In the following section, we shall use property (3) of theorem 1 to construct regular semigroups with (P).

2. Construction theorems

As was seen above, to construct regular semigroups with (P) it is needed only to describe every regular extension of $M \sim \Sigma\{M_\lambda : \lambda \in \Lambda\}$ by $Y(\Lambda)$ for a given H-degenerated inverse semigroup $Y(\Lambda)$ and for a given H-compatible completely regular semigroup $M \sim \Sigma\{M_\lambda : \lambda \in \Lambda\}$. It should be noted that structure theorems for H-compatible completely regular semigroups (that is, bands of groups) and for H-degenerated inverse semigroups have been already given by Schein [1973] and the author [1976] respectively.

Now, let $M \sim \Sigma\{M_\lambda : \lambda \in \Lambda\}$ be an H-compatible completely regular semigroup, and $Y(\Lambda)$ an H-degenerated inverse semigroup. A structure theorem for regular extensions of $M \sim \Sigma\{M_\lambda : \lambda \in \Lambda\}$ by $Y(\Lambda)$ can be obtained as a special case of the corollary to theorem 4 of [the author 1974], as follows (for the definition of a regular product, see [the author 1974]).

THEOREM 2. *Every regular extension of* $M \sim \Sigma\{M_\lambda: \lambda \in \Lambda\}$ *by* $Y(\Lambda)$ *can be obtained as a regular product of* M *and* $Y(\Lambda)$.

However, this construction is given in a very complicated form since the definition of a regular product is not brief. Therefore, we shall next consider the special case where $M \sim \Sigma\{M_\lambda: \lambda \in \Lambda\}$ is *upwards directed* (that is, the case where $E(M_\lambda)E(M_\tau) \subseteq E(M_\lambda)$ is satisfied for $\lambda, \tau \in \Lambda$ with $\lambda < \tau$). An orthodox union of groups, a completely simple semigroup and a primitive regular semigroup, etc., are of course upwards directed completely regular semigroups.

Let $M \sim \Sigma\{M_\lambda: \lambda \in \Lambda\}$ be upwards directed. Let u_λ be an idempotent of M_λ for each $\lambda \in \Lambda$. Suppose that S is a regular extension of $M \sim \Sigma\{M_\lambda: \lambda \in \Lambda\}$ by $Y(\Lambda)$. Then, there exists a surjective homomorphism $\phi: S \to Y(\Lambda)$ such that $\lambda\phi^{-1} = M_\lambda$ for each $\lambda \in \Lambda$. Put $\gamma\phi^{-1} = M_\gamma$ for $\gamma \in Y(\Lambda)\setminus\Lambda$, and take an element u_γ from each M_γ, $\gamma \in Y(\Lambda)\setminus\Lambda$. For each $\lambda \in \Lambda$, let L_λ be the L-class (of M_λ) containing u_λ, and R_λ the right zero semigroup of idempotents of the R-class (of M_λ) containing u_λ. Then, it is easy to see that the union $L[\Lambda]$ of all L_λ, that is, $L[\Lambda] = \cup\{L_\lambda: \lambda \in \Lambda\}$ and the union $R[\Lambda]$ of all R_λ, that is, $R[\Lambda] = \cup\{R_\lambda: \lambda \in \Lambda\}$ are a lower partial chain Λ of $\{L_\lambda: \lambda \in \Lambda\}$ and an upper partial chain Λ of $\{R_\lambda: \lambda \in \Lambda\}$ respectively (in the sense of [Warne 1976]) with respect to the multiplication in M. Further, it follows from [the author 1974] and [Warne 1976a] that any element x of $M_\gamma(\gamma \in Y(\Lambda))$ can be uniquely written in the form $x = au_\gamma e$, $a \in L_{\gamma\gamma^{-1}}$, $e \in R_{\gamma^{-1}\gamma}$. Further, let $x = au_\gamma e$ and $y = bu_\tau f$, where $a \in L_{\gamma\gamma^{-1}}$, $e \in R_{\gamma^{-1}\gamma}$, $b \in L_{\tau\tau^{-1}}$ and $f \in R_{\tau^{-1}\tau}$. Then, $xy = a(u_\gamma ebu_\tau)f$. Since $u_\gamma ebu_\tau \in M_{\gamma\tau}$, there exist $p \in L_{\gamma\tau(\gamma\tau)^{-1}}$ and $q \in R_{(\gamma\tau)^{-1}\gamma\tau}$ such that $u_\gamma ebu_\tau = pu_{\gamma\tau}q$. Hence, $xy = apu_{\gamma\tau}qf$. Now, we can consider the mappings $f_{\langle\gamma,\tau\rangle}: R_{\gamma^{-1}\gamma} \times L_{\tau\tau^{-1}} \to L_{\gamma\tau(\gamma\tau)^{-1}}$ and $g_{\langle\gamma,\tau\rangle}: R_{\gamma^{-1}\gamma} \times L_{\tau\tau^{-1}} \to R_{(\gamma\tau)^{-1}\gamma\tau}$ defined by $(e,b)f_{\langle\gamma,\tau\rangle} = p$ and $(e,b)g_{\langle\gamma,\tau\rangle} = q$. Then, the family $\Delta = \{f_{\langle\gamma,\tau\rangle}: \gamma,\tau \in Y(\Lambda)\} \cup \{g_{\langle\gamma,\tau\rangle}: \gamma,\tau \in Y(\Lambda)\}$ satisfies the following condition.

(C.1) (I) for $a \in L_{\gamma\gamma^{-1}}$, $e \in R_{\gamma^{-1}\gamma}$, $b \in L_{\delta\delta^{-1}}$, $f \in R_{\delta^{-1}\delta}$, $c \in L_{\tau\tau^{-1}}$ and $h \in R_{\tau^{-1}\tau}$,

$a(e,b((f,c)\, f_{\langle\delta,\tau\rangle}))f_{\langle\gamma,\delta\tau\rangle} = a((e,b)f_{\langle\gamma,\delta\rangle})((e,b)g_{\langle\gamma,\delta\rangle}f,c)f_{\langle\gamma\delta,\tau\rangle}$,

(II) for $a \in L_{\gamma\gamma^{-1}}$ and $e \in R_{\gamma^{-1}\gamma}$, there exist $b \in L_{\gamma^{-1}\gamma}$ and $f \in R_{\gamma\gamma^{-1}}$ such that $(e,b)f_{\langle\gamma,\gamma^{-1}\rangle}(f,a)f_{\langle\gamma\gamma^{-1},\gamma\rangle} \in E(L_{\gamma\gamma^{-1}})$,

(III) for $\lambda,\tau \in \Lambda$, $u \in R_\lambda$ and $v \in L_\tau$,

$uv = (u,v)f_{\langle\lambda,\tau\rangle}u_{\lambda\tau}(u,v)g_{\langle\lambda,\tau\rangle}$.

The product xy of x and y, where $x = au_\gamma e$, $y = bu_\tau f$, $a \in L_{\gamma\gamma^{-1}}$, $e \in R_{\gamma^{-1}\gamma}$, $b \in L_{\tau\tau^{-1}}$ and $f \in R_{\tau^{-1}\tau}$, can be given by

$$xy = a(e,b)f_{\langle\gamma,\tau\rangle}u_{\gamma\tau}(e,b)g_{\langle\gamma,\tau\rangle}f.$$

Hence, if we define multiplication in $\overline{S} = \{(a,\gamma,e)\colon a \in L_{\gamma\gamma^{-1}},$ $e \in R_{\gamma^{-1}\gamma},\ \gamma \in Y(\Lambda)\}$ by

(C.2) $(a,\gamma,e)(b,\tau,f) = (a(e,b)f_{\langle\gamma,\tau\rangle},\gamma\tau,(e,b)g_{\langle\gamma,\tau\rangle}f)$, then $\phi\colon \overline{S} \to S$ defined by $(a,\gamma,e)\phi = au_\gamma e$ is an isomorphism.

If a family $\Delta = \{f_{\langle\gamma,\tau\rangle}\colon \gamma,\tau \in Y(\Lambda)\} \cup \{g_{\langle\gamma,\tau\rangle}\colon \gamma,\tau \in Y(\Lambda)\}$ of mappings $f_{\langle\gamma,\tau\rangle}\colon R_{\gamma^{-1}\gamma} \times L_{\tau\tau^{-1}} \to L_{\gamma\tau(\gamma\tau)^{-1}}$ and $g_{\langle\gamma,\tau\rangle}\colon R_{\gamma^{-1}\gamma} \times L_{\tau\tau^{-1}} \to R_{(\gamma\tau)^{-1}\gamma\tau}$ satisfies (C.1), then we shall call Δ *a factor set of* $\{L[\Lambda], R[\Lambda]\}$ *belonging to* $Y(\Lambda)$.

Conversely, let $\Delta = \{f_{\langle\gamma,\tau\rangle}\colon \gamma,\tau \in Y(\Lambda)\} \cup \{g_{\langle\gamma,\tau\rangle}\colon \gamma,\tau \in Y(\Lambda)\}$ be any factor set of $\{L[\Lambda],R[\Lambda]\}$ belonging to $Y(\Lambda)$. Then, it is easily seen that $\overline{S} = \{(a,\gamma,e)\colon a \in L_{\gamma\gamma^{-1}},\ e \in R_{\gamma^{-1}\gamma},\ \gamma \in Y(\Lambda)\}$ becomes a regular extension of $M \sim \Sigma\{M_\lambda\colon \lambda \in \Lambda\}$ by $Y(\Lambda)$, up to isomorphism, under the multiplication given by (C.2). We denote this $\overline{S}$ by $R(L[\Lambda] \times Y(\Lambda) \times R[\Lambda];\Delta)$.

Thus, we have the following result.

THEOREM 3. *If $M \sim \Sigma\{M_\lambda\colon \lambda \in \Lambda\}$ is upwards directed, then any regular extension of $M \sim \Sigma\{M_\lambda\colon \lambda \in \Lambda\}$ by $Y(\Lambda)$ is isomorphic to some $R(L[\Lambda] \times Y(\Lambda) \times R[\Lambda];\Delta)$. Conversely, any $R(L[\Lambda] \times Y(\Lambda) \times R[\Lambda];\Delta)$ is a regular extension of $M \sim \Sigma\{M_\lambda\colon \lambda \in \Lambda\}$ by $Y(\Lambda)$, up to isomorphism.*

3. Orthodox semigroups with (*P*)

Let $Y(\Lambda)$ be an H-degenerated inverse semigroup, and $M \sim \Sigma\{M_\lambda : \lambda \in \Lambda\}$ an H-compatible completely regular semigroup. If $M \sim \Sigma\{M_\lambda : \lambda \in \Lambda\}$ is orthodox, then any regular extension of $M \sim \Sigma\{M_\lambda : \lambda \in \Lambda\}$ by $Y(\Lambda)$ is of course an orthodox semigroup with (P). Conversely, if a regular extension of $M \sim \Sigma\{M_\lambda : \lambda \in \Lambda\}$ by $Y(\Lambda)$ is orthodox then $M \sim \Sigma\{M_\lambda : \lambda \in \Lambda\}$ must be orthodox. On the other hand, it is easy to see from the definition of a strictly inversive semigroup (see [the author 1964]) that an H-compatible completely regular semigroup is orthodox if and only if it is a strictly inversive semigroup. As we shall deal with the construction of orthodox semigroups with (P), hereafter we assume that $M \sim \Sigma\{M_\lambda : \lambda \in \Lambda\}$ is a strictly inversive semigroup. Now, we shall give the definition of the spined product of two orthodox semigroups with (P). Let A and B be orthodox semigroups with (P) which have the same semigroup Γ as their greatest H-degenerated inverse semigroup homomorphic images. Let $f_1 : A \to \Gamma$ and $f_2 : B \to \Gamma$ be surjective homomorphisms. Put $\gamma f_1^{-1} = A_\gamma$ and $\gamma f_2^{-1} = B_\gamma$ for each $\gamma \in \Gamma$. Then, $C = \Sigma\{A_\gamma \times B_\gamma : \gamma \in \Gamma\}$ (where Σ denotes disjoint union) becomes an orthodox semigroup with (P) under the multiplication defined by

$$(a_\gamma, b_\gamma)(a_\tau, b_\tau) = (a_\gamma a_\tau, b_\gamma b_\tau),$$

where $(a_\gamma, b_\gamma) \in A_\gamma \times B_\gamma$ and $(a_\tau, b_\tau) \in A_\tau \times B_\tau$. We call C the spined product of A and B with respect to $(\Gamma; f_1, f_2)$, and denote it by $(A \bowtie B; \Gamma, f_1, f_2)$.

Since the semigroup M above is strictly inversive, it follows from [the author 1964] that M is isomorphic to the spined product of the band $E(M) \sim \Sigma\{E(M_\lambda) : \lambda \in \Lambda\}$ and a semilattice Λ of groups $\{G_\lambda : \lambda \in \Lambda\}$, say $C \sim \Sigma\{G_\lambda : \lambda \in \Lambda\}$, with respect to $\{\Lambda, \xi_1, \xi_2\}$, where ξ_1, ξ_2 are the homomorphisms of $E(M)$ and C onto Λ respectively defined by $e\xi_1 = \lambda$ for $e \in E(M_\lambda)$ and $a\xi_2 = \lambda$ for $a \in G_\lambda$. On the other hand, it is easy to see that any regular extension of $E(M) \sim \Sigma\{E(M_\lambda) : \lambda \in \Lambda\}$ by $Y(\Lambda)$ is an H-degenerated orthodox semigroup, while any regular extension of $C \sim \Sigma\{G_\lambda : \lambda \in \Lambda\}$ by $Y(\Lambda)$ is an H-compatible inverse semigroup. Therefore, we can conjecture that any regular extension of $M \sim \Sigma\{M_\lambda : \lambda \in \Lambda\}$ by $Y(\Lambda)$ is isomorphic to some spined product of an H-degenerated orthodox semigroup and an H-compatible inverse semigroup. The following theorem shows that this conjecture is actually true.

THEOREM 4. *Any regular extension S of a strictly inversive semigroup $M \sim \Sigma\{M_\lambda : \lambda \in \Lambda\}$ by an H-degenerated inverse semigroup $Y(\Lambda)$ is isomorphic to some spined product of an H-degenerated orthodox semigroup and an H-compatible inverse semigroup. Conversely, any spined product of an H-degenerated orthodox semigroup and an H-compatible inverse semigroup is an orthodox semigroup with (P).*

Proof. Let S be a regular extension of a strictly inversive semigroup $M \sim \Sigma\{M_\lambda : \lambda \in \Lambda\}$ by an H-degenerated inverse semigroup $Y(\Lambda)$. Then, there exists a surjective homomorphism $\phi: S \to Y(\Lambda)$ such that $\lambda\phi^{-1} = M_\lambda$ for each $\lambda \in \Lambda$. Let σ_ϕ be the congruence on S induced by ϕ, that is, $x\,\sigma_\phi\,y$ if and only if $x\phi = y\phi$.

Define further two relations ζ_S and η_S on S as follows:

(C.3) $a\,\zeta_S\,b$ if and only if $a\,\sigma_\phi\,b$ and $ab^*, b^*a \in E(S)$ for some $b^* \in V(b)$ (where $V(b)$ denotes the set of inverses of b).

(C.4) $a\,\eta_S\,b$ if and only if $a\,\sigma_\phi\,b$ and $aea^*beb^* = beb^*$, $beb^*aea^* = aea^*$ for all $e \in E(S)^1$ and for some $a^* \in V(a)$ and $b^* \in V(b)$; $a^*eab^*eb = a^*ea$, $b^*eba^*ea = b^*eb$ for all $e \in E(S)^1$ and for some $a^* \in V(a)$ and $b^* \in V(b)$.

Then, it is easily seen that ζ_S and η_S are congruences on S, and σ_ϕ, ζ_S and η_S satisfy the following: (1) $\zeta_S \subseteq \sigma_\phi$, $\eta_S \subseteq \sigma_\phi$, and ζ_S and η_S are permutable; (2) $\zeta_S \cap \eta_S = \iota_S$ (the identity congruence on S); and (3) $\zeta_S \vee \eta_S = \sigma_\phi$ (where $\zeta_S \vee \eta_S$ denotes the least congruence on S containing both ζ_S and η_S). Hence S is isomorphic to $(S/\zeta_S \bowtie S/\eta_S; S/\sigma_\phi, \phi_1, \phi_2)$, where ϕ_1 and ϕ_2 are homomorphisms of S/ζ_S and S/η_S, respectively, into S/σ_ϕ defined by $(x\zeta_S)\phi_1 = x\sigma_\phi$ and $(x\eta_S)\phi_2 = x\sigma_\phi$. Since $S/\sigma_\phi \simeq Y(\Lambda)$, under $\overline{\phi}$, S is isomorphic to $(S/\zeta_S \bowtie S/\eta_S; Y(\Lambda), \overline{\phi}_1, \overline{\phi}_2)$, where $\overline{\phi}$, $\overline{\phi}_1$ and $\overline{\phi}_2$ are the homomorphisms of S/σ_ϕ, S/ζ_S and S/η_S onto $Y(\Lambda)$ respectively defined by $(x\sigma_\phi)\overline{\phi} = x\phi$, $\overline{\phi}_1 = \phi_1\overline{\phi}$ and $\overline{\phi}_2 = \phi_2\overline{\phi}$. On the other hand, it is easily verified that S/ζ_S is an H-compatible inverse semigroup and S/η_S is an H-degenerated orthodox semigroup. Hence, the first half of the theorem is proved. The latter half is obvious.

Finally, it should also be noted that constructions for H-compatible inverse semigroups and H-degenerated orthodox semigroups have been given by the author [1976b].

References

1941 A.H. Clifford, Semigroups admitting relative inverses. *Ann. of Math.* 42 (1941), 1037-1049.

1964 M. Yamada, Strictly inversive semigroups. *Bull. Shimane Univ.* 13 (1964), 128-138.

1967a G. Lallement, Demi-groupes réguliers. *Ann. Mat. pura ed Appl.* 77 (1967), 47-129.

1967b M. Petrich, *Topics in Semigroups*. Pennsylvania State University, 1967.

1967c M. Yamada, Regular semigroups whose idempotents satisfy permutation identities. *Pacific J. Math.* 21 (1967), 371-392.

1969 T.E. Hall, On regular semigroups whose idempotents form a subsemigroup. *Bull. Australian Math. Soc.* 1 (1969), 195-208.

1970a T.E. Hall, On orthodox semigroups and uniform and antiuniform bands. *J. Algebra* 16 (1970), 204-217.

1970b M. Yamada, On a regular semigroup in which the idempotents form a band. *Pacific J. Math.* 33 (1970), 261-272.

1971 T.E. Hall, Orthodox semigroups. *Pacific J. Math.* 39 (1971), 677-686.

1973 B.M. Schein, Bands of unipotent monoids. *Semigroup Forum* 6 (1973), 75-79.

1974 M. Yamada, On regular extensions of a semigroup which is a semilattice of completely simple semigroups. *Mem. Fac. Lit & Sci., Shimane Univ., Nat. Sci.* 7 (1974), 1-17.

1976a R.J. Warne, Natural regular semigroups. *Colloq. Math. Soc. János Bolyai* 20 (1976), 685-720.

1976b M. Yamada, H-compatible orthodox semigroups. *Colloq. Math. Soc. János Bolyai* 20 (1976), 721-748.

TO BE PUBLISHED

[A] M. Yamada, On a certain class of regular semigroups.

INTERPOLATION ON SEMILATTICES

P. A. Grossman and H. Lausch

Department of Mathematics, Monash University, Clayton, Victoria, Australia, 3168.

Abstract

Let $\underline{\underline{V}}$ be a variety, A an algebra in $\underline{\underline{V}}$, and $n > 0$ an integer; $A^{(A^n)}$ is an algebra in $\underline{\underline{V}}$. Let $P_n(A)$ be the subalgebra of $A^{(A^n)}$ generated by the n projections from A^n to A and the constants; then $P_n(A)$ is called the algebra of n-place polynomial functions on A. A map ϕ from A^n to A is said to be a k-local polynomial map if, for any k elements $a_1,\ldots,a_k \in A^n$, there exists a polynomial function p such that $p(a_i) = \phi(a_i)$, $i = 1,\ldots,k$. The set of k-local polynomial maps from A^n to A is denoted by $L_kP_n(A)$. The behaviour of the chain $L_1P_n(A) \supseteq L_2P_n(A) \supseteq \ldots \supseteq L_kP_n(A) \supseteq \ldots$ has been investigated by various authors for a number of varieties, e.g. for any abelian group A and any n, $L_4P_n(A) = L_kP_n(A)$, for all $k \geqslant 4$, and $L_3P_1(A) = L_kP_1(A)$ for all $k \geqslant 3$ (Hule and Nöbauer [1977]). It will be shown that, for any semilattice $S, L_{n+2}P_n(S) = L_kP_n(S)$, for $k \geqslant n + 2$.

1. Interpolation on algebras

Let A be an algebra and $n \geqslant 1$ an integer. Then the set of all functions from A^n to A endowed with the operations of A defined "pointwise" constitutes an algebra $F_n(A)$ belonging to the variety generated by A, and is usually called the full n-place function algebra on A. The n projections $\xi_1,\ldots,\xi_n$ from A^n to A together with the constant functions from A^n to A generate a subalgebra $P_n(A)$ of $F_n(A)$, called the algebra of n-place polynomial functions on A [Lausch and Nöbauer 1973]. The usual polynomial functions in n "variables" over a ring are a special case of this construction.

An interpolation problem occurs when we ask whether, for a given function $\phi \in F_n(A)$ and a fixed positive integer k, it is always possible to find, for any k elements of A^n, a polynomial function $p \in P_n(A)$ such that p and ϕ coincide on the set of these k elements. Accordingly

ISBN 0-12-319450-4

we define $L_kP_n(A) = \{\phi \in F_n(A) |$ for any k elements $a_1, \ldots, a_k \in A^n$, there exists $p \in P_n(A)$ such that $\phi(a_i) = p(a_i)$, for $i = 1, \ldots, k\}$. It is easy to see that $L_kP_n(A)$ is a subalgebra of $F_n(A)$, and that $L_1P_n(A) = F_n(A)$. Furthermore we define $LP_n(A) = \bigcap_{k=1}^{\infty} L_kP_n(A)$, and call this algebra the algebra of local n-place polynomial functions. If A is finite, then $LP_n(A) = P_n(A)$. We thus obtain a descending chain of subalgebras of $F_n(A)$: $L_1P_n(A) \supseteq L_2P_n(A) \supseteq \ldots \supseteq LP_n(A) \supseteq P_n(A)$ which, in a way, describes interpolation of functions of $F_n(A)$.

Certain classes of algebras are defined by the property that certain members of this chain coincide, e.g. A is called n-(locally) polynomially complete if $F_n(A) = P_n(A)(LP_n(A))$. Such algebras have been studied by various authors, e.g. Fried, Kaiser and Márki [B]. A classical result due to Fröhlich [1958] states that any finite simple non-abelian group G is 1-polynomially complete. More generally, one can show that amongst the finite groups, the simple non-abelian groups are characterized by the property that $F_n(G) = P_n(G)$, for all $n \geqslant 1$ [Lausch and Nöbauer 1973].

In connexion with these results, Kollár [1979a] proved: It is not true that every semigroup S satisfying $F_n(S) = LP_n(S)$, for all $n \geqslant 1$, is necessarily a simple group; however, every semigroup S can be embedded into a semigroup $R(S)$ satisfying $F_n(R(S)) = LP_n(R(S))$, for all $n \geqslant 1$; if the given semigroup is a monoid, then the embedding can be chosen to be identity-preserving. Hule and Nöbauer [1977] proved that any abelian group A satisfies $L_3P_1(A) = LP_1(A)$ and $L_4P_n(A) = LP_n(A)$, for $n \geqslant 2$. Dorninger and Nöbauer [A] showed that for any lattice Λ, $L_2P_n(\Lambda) = LP_n(\Lambda)$, for all $n \geqslant 1$. Commutative rings with identity show, in general, a completely different behaviour: for the ring Z of integers, any two members of the abovementioned chain are distinct, whereas for factor rings of Z modulo a prime power p^e, there exist $k = k(p,e)$ such that all $L_iP_1(Z_{p^e})$ with $i \leqslant k$, are pairwise distinct and $L_kP_1(Z_{p^e}) = P_1(Z_{p^e})$ [Lausch and Nöbauer 1979b].

The objective of this paper is to study the interpolation behaviour of semilattices S: whereas for lattices the argument of Dorninger and Nöbauer [A] is a matter of a few lines, the absence of the other lattice operation $\cup$ on semilattices poses a number of problems. We show that $L_{n+2}P_n(S) = LP_n(S)$, for all $n \geqslant 1$ and for all semilattices S. So far we have been able to establish only for $n = 1$ that this is the best possible result.

2. Interpolation on semilattices

The laws for the variety of semilattices imply that every n-place polynomial function on a semilattice S can always be written in the form k, $\prod_{i\in I} \xi_i$, or $k \prod_{i\in I} \xi_i$, where $k \in S$, ξ_i is the i-th projection from S^n to S and I is some non-empty subset of $\{1,\ldots,n\}$. Throughout this section, S will always denote a semilattice.

PROPOSITION 1. *For each of (i) - (iii), a semilattice S can be found such that the set-theoretical inequalities below hold in the strict sense:*

(i) $L_1P_n(S) \supset L_2P_n(S)$;

(ii) $L_2P_n(S) \supset L_3P_n(S)$;

(iii) $LP_n(S) \supset P_n(S)$.

Proof. In each case it is sufficient to give an example for $n = 1$. (i) Let $S = \{a,b\}$, $a > b$, $g\colon S \to S$ be defined by $g(a) = b$, $g(b) = a$. Then $g \in L_1P_1(S)$, but $g \neq \xi_1$, $a\xi_1$, $b\xi_1$, hence $g \notin L_2P_1(S)$. (ii) Let $S = \{a,b,c,d\}$, $a,b,c > d$, $g\colon S \to S$ be defined by $g(a) = d$, $g(b) = b$, $g(c) = c$, $g(d) = d$. Then $b\xi_1, c\xi_1, d, \xi_1, \xi_1, \xi_1$ interpolate g at $\{a,b\},\{a,c\},\{a,d\},\{b,c\},\{b,d\},\{c,d\}$, respectively, hence $g \in L_2P_1(S)$; but g cannot be interpolated at $\{a,b,c\}$, as $\{b,c\}$ requires ξ_1 for interpolation whereas $g(a) = d \neq \xi_1(a)$. Hence $g \notin L_3P_1(S)$. (iii) Let $S = \{b,a_1,a_2,a_3,\ldots\}$, $b > a_1$, $a_{i+1} > a_i$, $i = 1,2,3,\ldots$, $g\colon S \to S$ be defined by $g(a_i) = a_i$, $i = 1,2,3,\ldots$, $g(b) = a_1$. On $\{b,a_1,\ldots,a_k\}$, for arbitrary k, $a_k\xi_1$ interpolates g, hence $g \in LP_1(S)$. But $g \neq \xi_1$ or $b\xi_1$, as $g(b) \neq b$; also $g \neq a_\ell\xi_1$, for any ℓ, as $g(a_{\ell+1}) \neq a_\ell$; nor is g constant. Hence $g \notin P_1(S)$.

PROPOSITION 2. *Any element $g \in L_3P_n(S)$ is a homomorphism from S^n to S.*

Proof. Let $g \in L_3P_n(S)$ and $r = (r_1,\ldots,r_n)$, $s = (s_1,\ldots,s_n)$ be elements in S^n. If the constant k interpolates g at r, s, and rs, then $g(rs) = k = k^2 = g(r)g(s)$. If $\prod_{i\in I} \xi_i$, $I \subseteq \{1,\ldots,n\}$, interpolates g at these three points, then $g(rs) = \prod_{i\in I} r_is_i = \prod_{i\in I} r_i \prod_{i\in I} s_i = g(r)g(s)$. If $k \prod_{i\in I} \xi_i$, $I \subseteq \{1,\ldots,n\}$, $k \in S$, interpolates g at r, s, and rs, $g(rs) = k \prod_{i\in I} r_i s_i = (k \prod_{i\in I} r_i)(k \prod_{i\in I} s_i) = g(r)g(s)$.

PROPOSITION 3. *If* $g \in L_2P_n(S)$, $s = (s_1,\dots,s_n) \in (g(S^n))^n$, *then there exists a non-empty set* $I \subseteq \{1,\dots,n\}$ *- which, in general, will depend on* s *- such that* $g(s) = \prod_{i\in I} s_i$.

Proof. Since $s_i \in g(S^n)$, $i = 1,\dots,n$, there exists $t(i) = (t(i)_1,\dots,t(i)_n) \in S^n$ such that $g(t(i)) = s_i$. For each i, let $p_i \in P_n(S)$ be an interpolating polynomial function for g at $t(i)$ and s. If for some i_0, p_{i_0} is constant, then $g(s) = g(t(i_0)) = s_{i_0}$, and the proposition follows in this case. If for some i_0, $p_{i_0} = \prod_{i\in I} \xi_i$, $I \subseteq \{1,\dots,n\}$, then $g(s) = \prod_{i\in I} s_i$ as claimed. Hence the case where $p_i = k_i \prod_{j\in I_i} \xi_j$, $I_i \subseteq \{1,\dots,n\}$, $1 \leqslant i \leqslant n$, remains. We then have

$$s_i = g(t(i)) = k_i \prod_{j\in I_i} t(i)_j, \tag{1}$$

$$g(s) = k_i \prod_{j\in I_i} s_j. \tag{2}$$

From (2) we obtain, for any pair i,ℓ:

$$k_i \prod_{j\in I_i} s_j = k_\ell \prod_{j\in I_\ell} s_j\ . \tag{3}$$

Once we have shown that $g(s) = \prod_{j\in I} s_j$, where $I = \bigcup_{i=1}^{n} I_i$, the proposition will follow. Define $T_i = \prod_{j\in I_i} t(i)_j$; then $s_i = k_i T_i$, by (1). Hence

$$\prod_{j\in I} s_j = \prod_{j\in I} k_j T_j = \prod_{j\in I} k_j k_j T_j = \prod_{j\in I} k_j \prod_{j\in I} k_j T_j = \prod_{j\in I} k_j \prod_{j\in I} s_j =$$

$$= \prod_{j\in I} k_j \prod_{m\in I_1} s_m \cdots \prod_{m\in I_n} s_m.$$

Rearranging the constants k_j, we obtain

$$\prod_{j\in I} s_j = [\prod_{j\in I} (k_j \prod_{m\in I_j} s_m)][\prod_{j\in I'} (\prod_{m\in I_j} s_m)] \tag{4}$$

where I' is the complement of I in $\{1,\dots,n\}$. Fix $j_0 \in I$; then by (3), $k_j \prod_{m\in I_j} s_m = k_{j_0} \prod_{m\in I_{j_0}} s_m$, for all $j \in I$, so $\prod_{j\in I}(k_j \prod_{m\in I_j} s_m) = k_{j_0} \prod_{m\in I_{j_0}} s_m$. Substituting in (4) yields:

$$\prod_{j\in I} s_j = k_{j_0} \prod_{m\in I_{j_0}} s_m [\prod_{j\in I'} (\prod_{m\in I_j} s_m)]\ . \tag{5}$$

Amongst all ways in which $\prod_{j\in I} s_j$ can be written in the form (5), choose one with $|I'|$ being minimal. We show $I' = \phi$, the empty set. If not, let $m_0 \in I'$; then (5) may be written:

$$\prod_{j\in I} s_j = k_{j_0} \prod_{m\in I_{j_0}} s_m \prod_{m\in I_{m_0}} s_m [\prod_{j\in I'\setminus\{m_0\}} (\prod_{m\in I_j} s_m)] . \tag{6}$$

By (3), $k_{j_0} \prod_{m\in I_{j_0}} s_m = k_{m_0} \prod_{m\in I_{m_0}} s_m$, and (6) becomes:

$$\prod_{j\in I} s_j = k_{m_0} \prod_{m\in I_{m_0}} s_m [\prod_{j\in I'\setminus\{m_0\}} (\prod_{m\in I_j} s_m)] . \tag{7}$$

(7) is of the same form as (5), with $I'\setminus\{m_0\}$ replacing I'. Minimality of I' implies $I' = \phi$ as claimed. Hence

$$\prod_{j\in I} s_j = k_{j_0} \prod_{m\in I_{j_0}} s_m . \tag{8}$$

Then (2) and (8) together imply $g(s) = \prod_{j\in I} s_j$ as asserted.

PROPOSITION 4. *If $g \in L_2P_n(S)$ and g is not a constant, then $g(S^n)$ is an ideal of S.*

Proof. Let $c \in g(S^n)$, and $C = (c,\ldots,c) \in S^n$. By proposition 3, $g(C) = c$. Let $\ell \in S$, $d = c\ell$, and $D = (d,\ldots,d) \in S^n$. We will show: $d \in g(S^n)$. Since g is not a constant, $g(D) \neq g(r)$, for some $r \in S^n$. Let $p \in P_n(S)$ interpolate g at r and D. p cannot be a constant, by the choice of r. If $p = \prod_{i\in I} \xi_i$, $I \subseteq \{1,\ldots,n\}$, then $g(D) = p(D) = d$, and we are done. Let $p = k \prod_{i\in I} \xi_i$, $k \in S$, $I \subseteq \{1,\ldots,n\}$; then $g(D) = p(D) = kd \leqslant d$. Let $q \in P_n(S)$ interpolate g at C and D. If $q = \prod_{j\in J} \xi_j$, $J \subseteq \{1,\ldots,n\}$, then $g(D) = q(D) = d$ whence $d \in g(S^n)$. If q is a constant, then $c = g(C) = g(D) \leqslant d \leqslant c$, hence $d = c \in g(S^n)$. If $q = m \prod_{j\in J} \xi_j$, $m \in S$, $J \subseteq \{1,\ldots,n\}$, then $g(C) = mc$ and $d \geqslant g(D) = md \geqslant mcd = g(C)d = cd = d$, hence $d = g(D) \in g(S^n)$, Q.E.D.

LEMMA 5. *Let M be a finite set, $|M| = n$, and let $A = \{A_\lambda\}_{\lambda\in\Lambda}$ be an indexed collection of distinct non-empty subsets of M such that $\lambda \neq \mu$ implies $A_\lambda \cap A_\mu = \cap A$. Then $|A| \leqslant n$.*

Proof. For each $\lambda \in \Lambda$, let $B_\lambda = A_\lambda \setminus \cap A$. Consider the collection $B = \{B_\lambda\}_{\lambda\in\Lambda} \cup \{\cap A\}$. It is readily checked that at most one of the sets in $\mathcal{B}$ is empty, and $\mathcal{B}$ is a collection of pairwise disjoint sets whose union is a subset of M. Hence the result follows.

PROPOSITION 6. *Let g be a function from S^n to S such that for any n elements $s(1),\dots,s(n) \in S^n$ there exists a subset $I \subseteq \{1,\dots,n\}$ such that $g(s(j)) = \prod_{i\in I} s(j)_i$, where $s(j) = (s(j)_1,\dots,s(j)_n)$, $j = 1,\dots,n$. Then, for any h elements $s(1),\dots,s(h) \in S^n$, $h \geqslant n$, there exists a subset $I' \subseteq \{1,\dots,n\}$ such that $g(s(j)) = \prod_{i\in I'} s(j)_i$, $j = 1,\dots,h$.*

Proof. By induction on h. For $h = n$, the proposition is true by assumption. Suppose the proposition holds for some $h \geqslant n$, and let $s(1),\dots,s(h+1) \in S^n$, $s(j) = (s(j)_1,\dots,s(j)_n)$, $j = 1,\dots,h+1$. By induction, for each i, $1 \leqslant i \leqslant h+1$, there exists a subset $L_i \subseteq \{1,\dots,n\}$ such that $g(s(j)) = \prod_{\lambda\in L_i} s(j)_\lambda$, for all $j \neq i$. For each i, we choose L_i such that $|L_i|$ is maximal for all possible choices of L_i. We observe that $\lambda \in L_i$ if and only if $g(s(j)) \leqslant s(j)_\lambda$ for all $j \neq i$. Hence if $\lambda \in L_i \cap L_k$, $i \neq k$, then $g(s(j)) \leqslant s(j)_\lambda$ for all $j \neq i$ and for all $j \neq k$, whence $g(s(j)) \leqslant s(j)_\lambda$ for all j, i.e. $\lambda \in L_i$, for $i = 1,\dots,h+1$. Hence any $\lambda \in \{1,\dots,n\}$ is either contained in all L_i, or in exactly one L_i, or in no L_i. We show that at most n distinct L_i can exist, satisfying this condition. Let, without loss of generality, $\{L'_1,\dots,L'_m\}$ be a maximal collection of pairwise distinct sets L'_i satisfying this condition. If for some λ, λ is in no L'_i, then we may add the set $\{\lambda\} \cup \{k \mid k \in L'_i,\ i = 1,\dots,m\}$ to our list of distinct sets L'_i to obtain a larger list with the same properties. Hence we may assume that $\bigcup_{i=1}^{m} L'_i = \{1,\dots,n\}$. We may now apply lemma 5 with $M = \{1,\dots,n\}$ and $A = \{L'_i\}_{i=1}^m$, to obtain $m \leqslant n$. Since $h + 1 > n$, there exist two distinct indices i and k, say, such that $1 \leqslant i < k \leqslant h+1$, and $L_i = L_k$. But then $g(s(j)) = \prod_{\lambda\in L_i} s(j)_\lambda$, for $j = 1,\dots,h+1$, Q.E.D.

PROPOSITION 7. *If $g \in L_{n+2}P_n(S)$ and there is a finite subset $F_0 \subseteq S^n$ for which there is no index set $I \subseteq \{1,\dots,n\}$ such that $g(s_1,\dots,s_n) = \prod_{i\in I} s_i$ for all $(s_1,\dots,s_n) \in F_0$, then $g(S^n)$ is directed upwards in S, i.e. if $e,f \in g(S^n)$, then there exists $c \in g(S^n)$ with $c \geqslant e$, $c \geqslant f$.*

Proof. Let $e,f \in g(S^n)$. If $e = f$, then the conclusion is trivial, so assume $e \neq f$. Let $p \in P_n(S)$ interpolate g at $E = (e,\dots,e)$, $F = (f,\dots,f)$, and $r(1),\dots,r(n) \in S^n$ which are chosen in such a way that for no subset $I \subseteq \{1,\ldots,n\}$, is $g(r(j)) = \prod_{i \in I} r(j)_i$, for each of $j = 1,\ldots,n$, where $r(j) = (r(j)_1,\ldots,r(j)_n)$. This choice is possible by the assumption on g and proposition 6. p cannot be constant as $p(E) = g(E) = e \neq f = g(F) = p(F)$, by proposition 3; nor can p be of the form $\prod_{i \in I} \xi_i$, for some $I \subseteq \{1,\ldots,n\}$, by the choice of the elements $r(j)$. Hence $p = k \prod_{i \in I} \xi_i$, for some $k \in S$ and $I \subseteq \{1,\ldots,n\}$. Consequently, $e = g(E) = ke$ and $f = g(F) = kf$. Let $K = (k,\ldots,k)$ and $c = g(K)$. Then $ce = g(K)g(E) = g(KE)$, by proposition 2. But $g(KE) = g(E) = e$, whence $ce = e$; i.e. $e \leqslant c$. Similarly $f \leqslant c$ follows.

PROPOSITION 8. *Let* $g \in L_{n+2}P_n(S)$ *and* $s(1),\ldots,s(h) \in (g(S^n))^n$. *Then* g *can be interpolated at these points by some* $p \in P_n(S)$ *with* $p = \prod_{i \in I} \xi_i$, *for some* $I \subseteq \{1,\ldots,n\}$.

Proof. By proposition 6, we may assume that $h = n$. If g does not satisfy the conditions of proposition 7, then the result is immediate. By proposition 7, we may therefore assume that $g(S^n)$ is directed upwards. Hence there exists $c \in g(S^n)$ such that $c \geqslant s(j)_i$, for $i,j = 1,\ldots,n$, where $s(j) = (s(j)_1,\ldots,s(j)_n)$ for $j = 1,\ldots,n$. By proposition 3, $g(s(j)) = \prod_{i \in I_j} s(j)_i$, for some $I_j \subseteq \{1,\ldots,n\}$. We consider two cases.

Case 1: $\prod_{i \in I_{j_0}} s(j_0)_i \neq c$, for some j_0. Let $p \in P_n(S)$ interpolate g at $s(1),\ldots,s(n)$, and $C = (c,\ldots,c)$. Then $g(C) = c$, by proposition 3, and $g(s(j_0)) = \prod_{i \in I_{j_0}} s(j_0)_i \neq c$, so p cannot be constant. If $p = \prod_{i \in I} \xi_i$ for some $I \subseteq \{1,\ldots,n\}$, the result is immediate. So we are left with $p = k \prod_{i \in I} \xi_i$, for some $k \in S$ and $I \subseteq \{1,\ldots,n\}$. In this case $c = g(C) = p(C) = kc$, so $c \leqslant k$. Therefore $s(j)_i \leqslant k$, for all $i,j = 1,\ldots,n$. Thus $g(s(j)) = p(s(j)) = k \prod_{i \in I} s(j)_i = \prod_{i \in I} s(j)_i$, and we are done.

Case 2: $\prod_{i \in I_j} s(j)_i = c$, for all j. Then $g(s(j)) = c$, for all j, so $c = \prod_{j=1}^{n} g(s(j)) = g(\prod_{j=1}^{n} s(j)_1,\ldots,\prod_{j=1}^{n} s(j)_n)$, by proposition 2. Since $s(j)_i \in g(S^n)$, for all i,j, and $g(S^n)$ is an ideal in S by proposition 4, we have $\prod_{j=1}^{n} s(j)_i \in g(S^n)$, for all j. Proposition 3 implies that $g(\prod_{j=1}^{n} s(j)_1,\ldots,\prod_{j=1}^{n} s(j)_n) = \prod_{i \in I} \prod_{j=1}^{n} s(j)_i$, for some $I \subseteq \{1,\ldots,n\}$, so $c = \prod_{i \in I} \prod_{i=1}^{n} s(j)_i$. Therefore, if $i \in I$, then $s(j)_i \geqslant c$, so $s(j)_i = c$, by the choice of c, for all j. Fix such an i in I, say

$i = i_0$. Then $g(s(j)) = c = s(j)_{i_0}$, for all j, i.e. $p = \xi_{i_0}$ is the required polynomial function.

THEOREM 9. *$L_{n+2}P_n(S) = LP_n(S)$, for $n = 1,2,3,\ldots$.*

Proof. Let $g \in L_{n+2}P_n(S)$. If g is constant, then $g \in LP_n(S)$. If g does not satisfy the conditions of proposition 7, then again $g \in LP_n(S)$. Assume therefore that g is not constant and satisfies the conditions of proposition 7. Let $r(1),\ldots,r(h) \in S^n$. Then $g(r(1)),\ldots,g(r(h)) \in g(S^n)$, and since by proposition 7, $g(S^n)$ is directed upwards in S, there exists $c \in g(S^n)$ such that $g(r(j)) \leqslant c$, for $j = 1,\ldots,h$. Let $C = (c,\ldots,c)$; then $g(C) = c$, by proposition 3. By proposition 2, $g(r(j)) = cg(r(j)) = g(C)g(r(j)) = g(Cr(j))$. By proposition 4, $g(S^n)$ is an ideal whence $cr(j)_i \in g(S^n)$, for all i,j, where $r(j) = (r(j)_1,\ldots,r(j)_n)$. Thus $Cr(j) \in (g(S^n))^n$, for $j = 1,\ldots,h$. By proposition 8, $g(Cr(j)) = \prod_{i\in I} cr(j)_i$, for $j = 1,\ldots,h$, and some $I \subseteq \{1,\ldots,n\}$, so $g(r(j)) = \prod_{i\in I} cr(j)_i = c\prod_{i\in I} r(j)_i$, i.e. $p = c\prod_{i\in I}\xi_i$ interpolates g at $r(1),\ldots,r(h)$. Hence $g \in LP_n(S)$.

ACKNOWLEDGEMENT. The authors are indebted to Dr C.J. Ash who provided them with proofs for proposition 4 and proposition 7 in the case of $n = 1$.

Final comment. After completion of the manuscript the first author proved the following result: If S is a semilattice then $L_3P_n(S) = LP_n(S)$ for all n.

References

1958 A. Fröhlich, The near-ring generated by the inner automorphisms of a finite simple group. *J. London Math. Soc.* 33 (1958), 95-107.

1973 H. Lausch and W. Nöbauer, *Algebra of polynomials.* North-Holland Publishing Company, Amsterdam, London, 1973.

1977 H. Hule and W. Nöbauer, Local polynomial functions on universal algebras. *Anais da Acad. Brasileira de Ciências* 49 (1977), 365-372.

1979a J. Kollár, Interpolation property in semigroups. *Semigroup Forum* 17 (1979), 337-350.

1979b H. Lausch and W. Nöbauer, Local polynomial functions on factor rings of the integers. *J. Austral. Math. Soc.* (Series A) 27 (1979), 232-238.

TO BE PUBLISHED

[A] D. Dorninger and W. Nöbauer, Local polynomial functions on lattices and universal algebras. *Coll. Math.*, to appear.

[B] E. Fried, H.K. Kaiser and L. Márki, Pattern functions and interpolation. To appear.

CONSTRUCTING BIORDERED SETS

John Meakin

Department of Mathematics and Statistics, The University of Nebraska–Lincoln,
Lincoln, Nebraska, U.S.A., 68588.

Abstract

Relative to the operation of forming "basic products" the set of idempotents of a regular semigroup forms a partial binary algebra which has been axiomatically characterized by K.S.S. Nambooripad as a "biordered set" (K.S.S. Nambooripad, *Structure of Regular Semigroups* I, Memoirs, Amer. Math. Soc. 244 (1979)). In this paper we indicate some techniques for constructing biordered sets from semilattices and rectangular biordered sets. In particular we describe a construction of all coextensions of rectangular biordered sets by semilattices, a construction of solid biordered sets and a construction of all pseudo-semilattices. We show how all finite biordered sets may be constructed from (finite) semilattices and (finite) rectangular biordered sets.

1. Biordered sets

Let E be the set of idempotents of a regular semigroup S. On E define relations ω^r and ω^ℓ as follows: for $e,f \in E$,

$$e\ \omega^r\ f \text{ if and only if } e = fe, \text{ and}$$

$$e\ \omega^\ell\ f \text{ if and only if } e = ef.$$

Let $\omega = \omega^r \cap \omega^\ell$ (the natural partial order on the idempotents of S), $\kappa = \omega^r \cup \omega^\ell$, $R = \omega^r \cap (\omega^r)^{-1}$ and $L = \omega^\ell \cap (\omega^\ell)^{-1}$. For $\rho \in \{\omega^\ell, \omega^r, \omega\}$ and $e \in E$ we let $\rho(e) = \{f \in E\colon f\ \rho\ e\}$. If $e\ \omega^r f$ it is easy to see that $ef \in E$, $ef\ R\ e$ and $ef\ \omega\ f$. In particular, the products ef and fe are defined in E. Similarly, if $e\ \omega^\ell\ f$, the products ef and fe are defined in E. Thus the product ef of $e,f \in E$ must be defined in E whenever $e(\kappa \cup \kappa^{-1})f$. Products of this form are referred to as *basic products*. Relative to these basic products, E forms a partial binary algebra with domain $D_E = \kappa \cup \kappa^{-1}$.

In [1979b], K.S.S. Nambooripad characterized the partial binary algebra E (relative to these basic products) axiomatically as a biordered

ISBN 0-12-319450-4

set. Nambooripad's axioms for biordered sets are provided in the following definition.

DEFINITION 1.1. Let E be a partial binary algebra. On E define relations ω^r, ω^ℓ, ω, κ, R and L as above (where we interpret the product ef for $e,f \in E$ as the product in the partial binary algebra E). Then E is a *biordered set* if the following axioms and their duals hold. (Here e,f,g,h etc. denote arbitrary elements of E.)

(B1) ω^r and ω^ℓ are quasi-orders on E and $D_E = \kappa \cup \kappa^{-1}$. ($D_E$ is the domain of the partial binary operation on E.)

(B21) $f \in \omega^r(e)$ implies $f \mathrel{R} fe \mathrel{\omega} e$.

(B22) $g \mathrel{\omega^\ell} f$, $f,g \in \omega^r(e)$ implies $ge \mathrel{\omega^\ell} fe$.

(B31) $g \mathrel{\omega^r} f \mathrel{\omega^r} e$ implies $(ge)f = gf$.

(B32) $g \mathrel{\omega^\ell} f$, $g,f \in \omega^r(e)$ implies $(fg)e = (fe)(ge)$.

(B4) $f,g \in \omega^r(e)$ implies $S(f,g)e = S(fe,ge)$.

Here $S(e,f)$ is defined for $e,f \in E$ as follows: first let $M(e,f)$ denote the quasi-ordered set

$$M(e,f) = (\omega^\ell(e) \cap \omega^r(f), \prec)$$

where $\prec$ is defined by

$$g \prec h \text{ (for } g,h \in M(e,f)) \text{ iff } eg \mathrel{\omega^r} eh \text{ and } gf \mathrel{\omega^\ell} hf.$$

Then $S(e,f) = \{h \in M(e,f)\colon g \prec h \text{ for all } g \in M(e,f)\}$ is called the *sandwich set* of e and f (in that order).

The biordered set E is said to be *regular* if

(R) $S(e,f) \neq \square$, for all $e,f \in E$.

We shall be concerned only with regular biordered sets in this paper: consequently we shall omit the adjective "regular" and consider a biordered set as a partial algebra satisfying (B1) - (B4) and their duals and (R).

DEFINITION 1.2. A *bimorphism* from the biordered set E to the biordered set F is a mapping $\theta\colon E \to F$ satisfying the following two properties:

(a) if $e,f \in E$ then $(e,f) \in D_E$ implies $(e\theta, f\theta) \in D_F$ and $(ef)\theta = (e\theta)(f\theta)$;

(b) $S(e,f)\theta \subseteq S(e\theta, f\theta)$.

We consider biordered sets as forming a category in which objects are biordered sets and morphisms are bimorphisms.

Using the concept of a biordered set, Nambooripad [1979b] has developed a theory of fundamental regular semigroups analogous to Munn's theory [1970] of fundamental inverse semigroups: he has also determined all regular partial bands and all regular idempotent-generated semigroups determined by a given biordered set. His theory rests solidly upon the notion of a biordered set and clearly demonstrates a need to study the structure of biordered sets. The axioms are somewhat complicated and it is desirable to study ways of building biordered sets from simpler kinds of structures.

The simplest kinds of biordered sets which readily come to mind are semilattices and rectangular biordered sets. A *semilattice* is a biordered set E in which $\omega^r = \omega^\ell = \omega$ (or equivalently $R = L = \iota_E$). It is easy to see that if $\omega^r = \omega^\ell$ then, for all $e,f \in E$, $|S(e,f)| = 1$ and $S(e,f) = S(f,e)$: if we define $e \wedge f = f \wedge e$ to be the unique element in $S(e,f)$ then $(E,\wedge)$ becomes a semilattice in the usual sense. Conversely if $(E,\wedge)$ is a semilattice we may think of E as a biordered set in which $\omega^r = \omega^\ell$. Thus we shall identify semilattices with biordered sets in which $\omega^r = \omega^\ell$.

A *rectangular biordered set* is a biordered set E such that for all $e,f \in E$ there exist $g,h \in E$ with

$$e \; R \; g \; L \; f \; R \; h \; L \; e.$$

Clearly a rectangular biordered set is the biordered set obtained from a rectangular band $I \times \Lambda$ by omitting all but the basic products (i.e. products of the form $(i,\lambda)(i,\mu) = (i,\mu)$ and of the form $(i,\lambda)(j,\lambda) = (i,\lambda)$). It is very easy to establish the following result.

PROPOSITION 1.3. *For a biordered set E the following are equivalent.*

(a) E is a rectangular biordered set.

(b) $\omega = \iota_E$.

(c) *E is the biordered set of some completely simple semigroup.*

We remark that from Nambooripad [1979b], it follows that if E is a biordered set and $e,f \in E$, then $S(e,f)$ is a rectangular biordered subset of E.

A natural question to ask seems to be the following: which biordered sets can be "built" from rectangular biordered sets and semilattices: what processes are available for building biordered sets from rectangular biordered sets and semilattices? Some building processes which have proved useful are the processes of taking coextensions and images by rectangular biordered sets and semilattices. We define these processes as follows.

DEFINITION 1.4. Let E and F be biordered sets. We say that E is a coextension of F by semilattices [rectangular biordered sets] or that F is an image of E by semilattices [rectangular biordered sets] if there is a bimorphism $\theta: E \twoheadrightarrow F$ from E onto F such that, for each $e \in E$, $e(\theta \circ \theta^{-1})$ is a semilattice [rectangular biordered set].

We are interested in constructing biordered sets from semilattices and rectangular biordered sets by repeated application of the processes of taking coextensions or images by rectangular biordered sets and semilattices.

2. Coextensions of rectangular biordered sets by semilattices

We proceed to describe a construction of all coextensions of rectangular biordered sets by semilattices. This construction appears in a paper of Byleen, Meakin and Pastijn [A] and has been developed and greatly exploited by Pastijn [I].

Let I and Λ be sets and let $\{L_\lambda : \lambda \in \Lambda\}$ and $\{R_i : i \in I\}$ be families of semilattices. Let $\{E_{i\lambda} : (i,\lambda) \in I \times \Lambda\}$ be a family of pairwise disjoint semilattices and let $\phi_{i\lambda}: E_{i\lambda} \to L_\lambda$ and $\psi_{i\lambda}: E_{i\lambda} \to R_i$ (for $(i,\lambda) \in I \times \Lambda$) be monomorphisms of $E_{i\lambda}$ onto *ideals* of L_λ and R_i respectively such that, for all (i,λ), $(j,\mu) \in I \times \Lambda$, all $x_{i\lambda} \in E_{i\lambda}$ and all $y_{j\mu} \in E_{j\mu}$,

$$(\langle x_{i\lambda}\psi_{i\lambda}\rangle \cap E_{i\mu}\psi_{i\mu})\psi_{i\mu}^{-1} \cap (\langle y_{j\mu}\phi_{j\mu}\rangle \cap E_{i\mu}\phi_{i\mu})\phi_{i\mu}^{-1} = \langle z_{i\mu}\rangle \qquad (*)$$

for some $z_{i\mu} \in E_{i\mu}$. (Here by $\langle x\rangle$ we mean the principal ideal generated by x in a semilattice.) On $E = \bigcup_{(i,\lambda)\in I\times\Lambda} E_{i\lambda}$ define relations ω^{ℓ} and ω^{r} as follows:

$$e_{i\lambda}\ \omega^{\ell}\ f_{j\mu} \text{ iff } \lambda = \mu \text{ and } e_{i\lambda}\ \phi_{i\lambda} \leqslant f_{j\mu}\ \phi_{j\mu} \text{ in } L_{\lambda},$$

$$e_{i\lambda}\ \omega^{r}\ f_{j\mu} \text{ iff } i = j \text{ and } e_{i\lambda}\ \psi_{i\lambda} \leqslant f_{j\mu}\ \psi_{j\mu} \text{ in } R_{i};$$

define basic products on E in the following way:

if $e_{i\lambda}\ \omega^{\ell}\ f_{j\mu}$ define $e_{i\lambda} f_{j\mu} = e_{i\lambda}$ and $f_{j\mu} e_{i\lambda} = e_{i\lambda}\ \phi_{i\lambda}\ \phi_{j\mu}^{-1}$;

if $e_{i\lambda}\ \omega^{r}\ f_{j\mu}$ define $e_{i\lambda} f_{j\mu} = e_{i\lambda}\ \psi_{i\lambda}\ \psi_{j\mu}^{-1}$ and $f_{j\mu} e_{i\lambda} = e_{i\lambda}$.

Denote the partial binary algebra E obtained this way by $E(L_{\lambda};R_{i};E_{i\lambda};\phi_{i\lambda};\psi_{i\lambda};I,\Lambda)$. From [A] and [I] we have the following theorem.

THEOREM 2.1. *The partial binary algebra* $E(L_{\lambda};R_{i};E_{i\lambda};\phi_{i\lambda};\psi_{i\lambda};I,\Lambda)$ *is a biordered set which is a coextension of the rectangular biordered set* $I \times \Lambda$ *by semilattices. Every biordered set which is a coextension of a rectangular biordered set by semilattices is obtained this way.*

We remark that condition (*) is automatically satisfied if each of the semilattices E_{i}, for $(i,\lambda) \in I \times \Lambda$, has an identity. Condition (*) is simply the condition which is necessary in order to guarantee that sandwich sets of pairs of elements in $E = \bigcup_{(i,\lambda)\in I\times\Lambda} E_{i\lambda}$ are non-empty.

The construction outlined above has proved useful in constructing numerous interesting examples of biordered sets. We briefly mention two examples here which have appeared in the literature.

EXAMPLE 2.2. Let $I = \Lambda = \{1,2\}$, let $R_1 = R_2 = L_1 = L_2 = \mathbb{Z}^{-}$ (the negative integers relative to the usual ordering) and for $i,j \in \{1,2\}$ let $E_{ij} = \{n_{ij}\colon n \in \mathbb{Z}^{-}\}$ with order $n_{ij} \geqslant m_{ij}$ iff $n \geqslant m$. For $n \in \mathbb{Z}^{-}$ let $\phi_{11} = \psi_{11}\colon n_{11} \mapsto n$, $\phi_{22} = \psi_{22}\colon n_{22} \mapsto n$, $\phi_{12} = \psi_{12}\colon n_{12} \mapsto n$, $\psi_{21}\colon n_{21} \mapsto n$ and $\phi_{21}\colon n_{21} \mapsto (n-1)_{21}$. The biordered set $E(L_{\lambda};R_{i};E_{i\lambda};\phi_{i\lambda};\psi_{i\lambda};I,\Lambda)$ is the four-spiral biordered set of Byleen, Meakin and Pastijn [1978]. It was the first known example of a connected biordered set which is not rectangular, and is in some sense the "smallest" such biordered set.

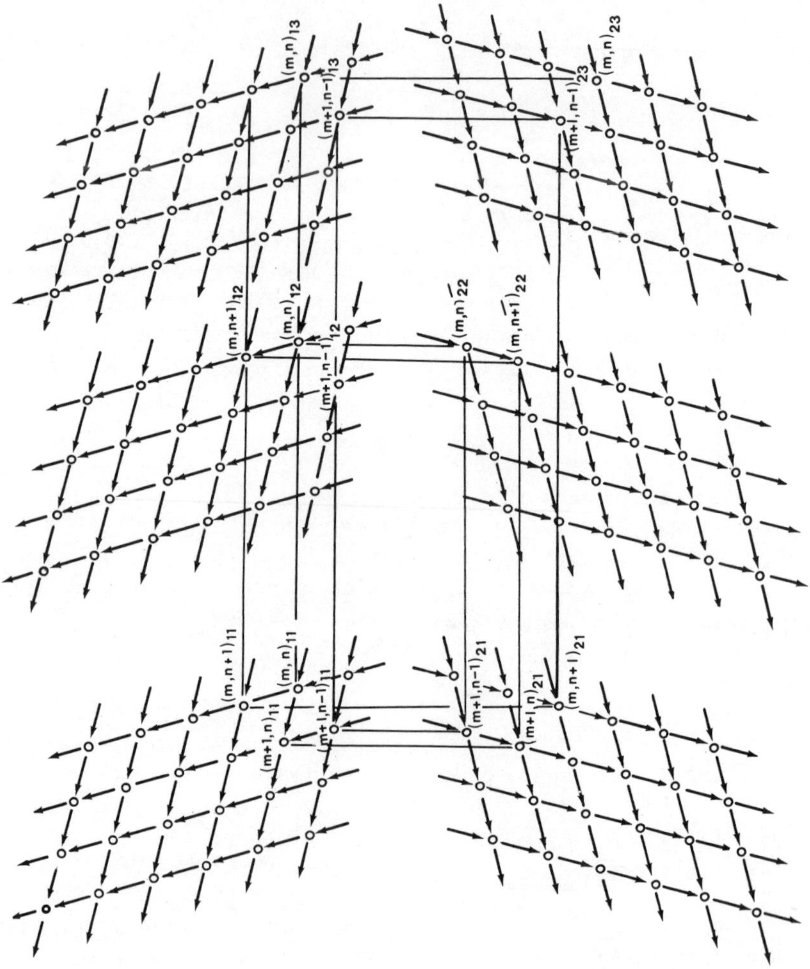
(m,n)13
(m+1,n-1)13
(m,n)23
(m+1,n-1)23
(m,n+1)12
(m,n)12
(m+1,n-1)12
(m,n)22
(m,n+1)22
(m,n+1)11
(m,n)11
(m+1,n-1)11
(m+1,n)11
(m+1,n-1)21
(m+1,n)21
(m,n+1)21

Diagram 1

EXAMPLE 2.3. Let $I = \{1,2\}$, $\Lambda = \{1,2,3\}$, let $R_1 = R_2 = L_1 = L_2 = L_3 = (\mathbb{Z} \times \mathbb{Z}, \leqslant)$ where $\mathbb{Z}$ denotes the set of integers and the order on $\mathbb{Z} \times \mathbb{Z}$ is defined by $(a,b) \leqslant (c,d)$ iff $a \geqslant c$ and $b \geqslant d$. For $(i,\lambda) \in I \times \Lambda$ let $E_{i\lambda} = \{(a,b)_{i\lambda}: (a,b) \in \mathbb{Z} \times \mathbb{Z}\}$ with order $(a,b)_{i\lambda} \leqslant (c,d)_{i\lambda}$ iff $a \geqslant c$ and $b \geqslant d$. Define

$$\psi_{22}: (m,n)_{22} \mapsto (m+1, n-1),$$

$$\psi_{23}: (m,n)_{23} \mapsto (m-1, n+2)$$

$$\psi_{i\lambda}: (m,n)_{i\lambda} \mapsto (m,n) \text{ if } (i,\lambda) \neq (2,2) \text{ or } (2,3) \text{ and}$$

$$\phi_{i\lambda}: (m,n)_{i\lambda} \mapsto (m,n) \text{ for all } (i,\lambda) \in I \times \Lambda.$$

The biordered set $E(L_\lambda; R_i; E_{i\lambda}; \phi_{i\lambda}; \psi_{i\lambda}; I, \Lambda)$ is pictured in diagram 1: it appears in the paper of Byleen, Meakin and Pastijn [A] as an example of a connected biordered set which is not rectangular and which does not contain a copy of the four-spiral biordered set.

3. Pseudo-semilattices

The biordered sets $E(L_\lambda; R_i; E_{i\lambda}; \phi_{i\lambda}; \psi_{i\lambda}; I, \Lambda)$ constructed in the previous section are all examples of pseudo-semilattices. In [1972], B.M. Schein defined a *pseudo-semilattice* to be a structure $(E, \omega^\ell, \omega^r)$ consisting of a set E and two quasi-orders ω^ℓ and ω^r on E such that to every pair $e,f \in E$ there exists a unique element $e \wedge f \in E$ for which

$$\omega^r(e) \cap \omega^\ell(f) = \omega(e \wedge f). \qquad (**)$$

(Here $\omega = \omega^\ell \cap \omega^r$ again, and for $e \in E$ and $\rho \in \{\omega^\ell, \omega^r, \omega\}$, $\rho(e) = \{f \in E: f \,\rho\, e\}$.) Actually Schein introduced the operation $\wedge$ in a dual fashion to the above: we dualize his definition in order that the operation $\wedge$ will extend the basic products in the context in which E is a biordered set.

In [H] K.S.S. Nambooripad studied the class of biordered sets which form pseudo-semilattices relative to the usual ω^ℓ and ω^r operations. He established that a biordered set E is a pseudo-semilattice if and only if $|S(e,f)| = 1$ for all $e,f \in E$: he further showed that if this is the case then the following conditions and their duals must hold:

for $e,f,g \in E$,

(PA1) if $e \ \omega^r \ f$ then $f \wedge e = e$;

(PA2) if $f,g \in \omega^r(e)$ then

(i) $(g \wedge e) \wedge f = g \wedge f$, and

(ii) $(g \wedge f) \wedge e = g \wedge (f \wedge e)$.

These conditions amount to the fact that for each $e \in E$, $(\omega^\ell(e),\wedge)$ and $(\omega^r(e),\wedge)$ are semigroups. Nambooripad referred to a pseudo-semilattice which satisfies (PA1), (PA2) and their duals as a *partially associative* pseudo-semilattice. In this paper *all pseudo-semilattices will be assumed to be partially associative*: consequently we shall omit the adjectives "partially associative" and refer to a *pseudo-semilattice as a structure* (E,ω^ℓ,ω^r) *satisfying* (**), (PA1), (PA2) *and their duals.*

If (E,ω^ℓ,ω^r) is a pseudo-semilattice and we define basic products by

$$ef = f \quad \text{and} \quad fe = f \wedge e \quad \text{when} \quad f \in \omega^r(e)$$

and

$$fe = f \quad \text{and} \quad ef = e \wedge f \quad \text{when} \quad f \in \omega^\ell(e),$$

then E becomes a biordered set with $S(e,f) = \{f \wedge e\}$ for all $e,f \in E$. The category of pseudo-semilattices (with morphisms mappings $\phi\colon E \to E'$ which preserve the $\wedge$ operation) is isomorphic to the category of biordered sets with trivial sandwich sets (in which morphisms are bimorphisms) and we shall identify these categories.

From Nambooripad's papers [H] and [G] we have the following result.

THEOREM 3.1. *The following conditions on a regular semigroup S are equivalent:*

(a) for all $e \in E(S)$, eSe is an inverse semigroup;

(b) $(E(S),\omega^\ell,\omega^r)$ is a pseudo-semilattice;

(c) for all $e \in E(S)$, $\omega(e)$ is a semilattice;

(d) for all $e \in E(S)$, $\omega^\ell(e)$ is a left normal band and $\omega^r(e)$ is a right normal band;

(e) for all $e,f \in E(S)$, $|S(e,f)| = 1$;

(f) the natural partial order on S *is compatible with the multiplication of* S *(see Nambooripad [12]),*

(g) if $e,f,g \in E(S)$, $f,g \in \omega(e)$ *and* $f(\mathcal{R} \cup \mathcal{L})g$, *then* $f = g$.

A regular semigroup which satisfies any one of the equivalent conditions of Theorem 3.1 is called a *pseudo-inverse semigroup*. Examples of such semigroups abound: for example completely 0-simple semigroups, generalized inverse semigroups (Yamada [1967]), locally testable regular semigroup (Zalcstein [1973b]) are all pseudo-inverse. Any regular semigroup (such as the fundamental four-spiral semigroup [1978]) whose biordered set is of the form $E(L_\lambda;R_i;E_{i\lambda};\phi_{i\lambda};\psi_{i\lambda};I,\Lambda)$ is pseudo-inverse.

In [E], Meakin and Pastijn have obtained a construction of all pseudo-semilattices from rectangular biordered sets and semilattices. We have the following theorem.

THEOREM 3.2. *If* E *is a biordered set which is an image by rectangular biordered sets of a biordered set of the form* $E(L_\lambda;R_i;E_{i\lambda};\phi_{i\lambda};\psi_{i\lambda};I,\Lambda)$, *then* E *is a pseudo-semilattice: every pseudo-semilattice* E *is an image by rectangular biordered sets of a biordered set of the form* $E(L_\lambda;R_i;E_{i\lambda};\phi_{i\lambda};\psi_{i\lambda};I,\Lambda)$.

We shall outline a proof of this theorem based on the following construction which is a slight generalization of a construction given by Meakin and Pastijn [E] in their proof of this theorem.

CONSTRUCTION 3.3. *Let* E *be a biordered set and let*

$$\overline{E} = \{(e,g,f) \in E \times E \times E\colon g \in M(f,e)\}.$$

On $\overline{E}$ *define*

$$(e_1,g_1,f_1)\ \omega^r\ (e_2,g_2,f_2) \text{ iff } e_1 = e_2 \text{ and } g_1\ \omega^r\ g_2,$$

$$(e_1,g_1,f_1)\ \omega^\ell\ (e_2,g_2,f_2) \text{ iff } f_1 = f_2 \text{ and } g_1\ \omega^\ell\ g_2$$

and define

$$(e_1,g_1,f_1)(e_2,g_2,f_2) = (e_1,g_1g_2,f_2)$$

whenever $(e_1,g_1,f_1)(\kappa \cup \kappa^{-1})\ (e_2,g_2,f_2)$. *Then* $\overline{E}$ *is a biordered set* *which is a coextension of E by rectangular biordered sets: the mapping $\theta: \overline{E} \twoheadrightarrow E \times E$ defined by $(e,g,f)\theta = (e,f)$ is a bimorphism from $\overline{E}$ onto the rectangular biordered set $E \times E$ and for each $(e,f) \in E \times E$, $(e,f)\theta^{-1}$ is a biordered subset of $\overline{E}$ biorder isomorphic to $M(f,e)$.*

Proof. This can be proved directly, using the axioms for biordered sets: a somewhat simpler proof that $\overline{E}$ is a biordered set is as follows. Let S be any regular semigroup with $E(S) = E$ and let $\overline{S} = \{(e,x,f) \in E \times S \times E: x \in eSf\}$: on $\overline{S}$ define a multiplication by $(e,x,f)(g,y,h) = (e,xy,h)$. It is routine to check that $\overline{S}$ is a regular semigroup with $E(\overline{S}) = \overline{E}$ and it follows that $\overline{E}$ is a biordered set. It is immediate from the definition of bimorphism that θ is a bimorphism from $\overline{E}$ onto the rectangular biordered set $E \times E$ with the stated property. Denote the projection from $\overline{E}$ onto E defined by $(e,g,f) \mapsto g$ by π: clearly π is a bimorphism from $\overline{E}$ onto E and for $g \in E$, $g\pi^{-1} = \{(e,g,f): g \in M(f,e)\}$ is a rectangular biordered set since the ω-relation on the biordered subset $g\pi^{-1}$ of $\overline{E}$ is trivial. This completes the proof of 3.3.

Proof of theorem 3.2. It is easy to see that principal ideals of a biordered set of the form $F = E(L_\lambda;R_i;E_{i\lambda};\phi_{i\lambda};\psi_{i\lambda};I,\Lambda)$ are semilattices and so F is a pseudo-semilattice: it follows that any bimorphic image of F is also a pseudo-semilattice (see Nambooripad [H]). Conversely, if E is any pseudo-semilattice, let $\overline{E}$ be the biordered set constructed from E as in construction 3.3: since $M(f,e) = \omega(e \wedge f)$ is a semilattice for all $e,f \in E$, it follows that $\overline{E}$ is a coextension of the rectangular biordered set $E \times E$ by semilattices and hence $\overline{E} \cong E(L_\lambda;R_i;E_{i\lambda};\phi_{i\lambda};\psi_{i\lambda};I,\Lambda)$ for suitable choice of $L_\lambda, R_i, E_{i\lambda}, \phi_{i\lambda}, \psi_{i\lambda}$, I and Λ. The result of theorem 3.2 now follows immediately from construction 3.3.

In [H] Nambooripad showed that pseudo-semilattices may be regarded as binary algebras $(E,\wedge)$ where the operation $\wedge$ satisfies the following identities and their duals: for all $x,y,z,u \in E$

(a) $x \wedge x = x$

(b) $(x \wedge y) \wedge (x \wedge z) = (x \wedge y) \wedge z$

(c) $(x \wedge y) \wedge ((x \wedge z) \wedge (x \wedge u)) = ((x \wedge y) \wedge (x \wedge z)) \wedge (x \wedge u)$.

The category of pseudo-semilattices is isomorphic to the category of binary algebras $(E,\wedge)$ satisfying the identities (a), (b) and (c) and

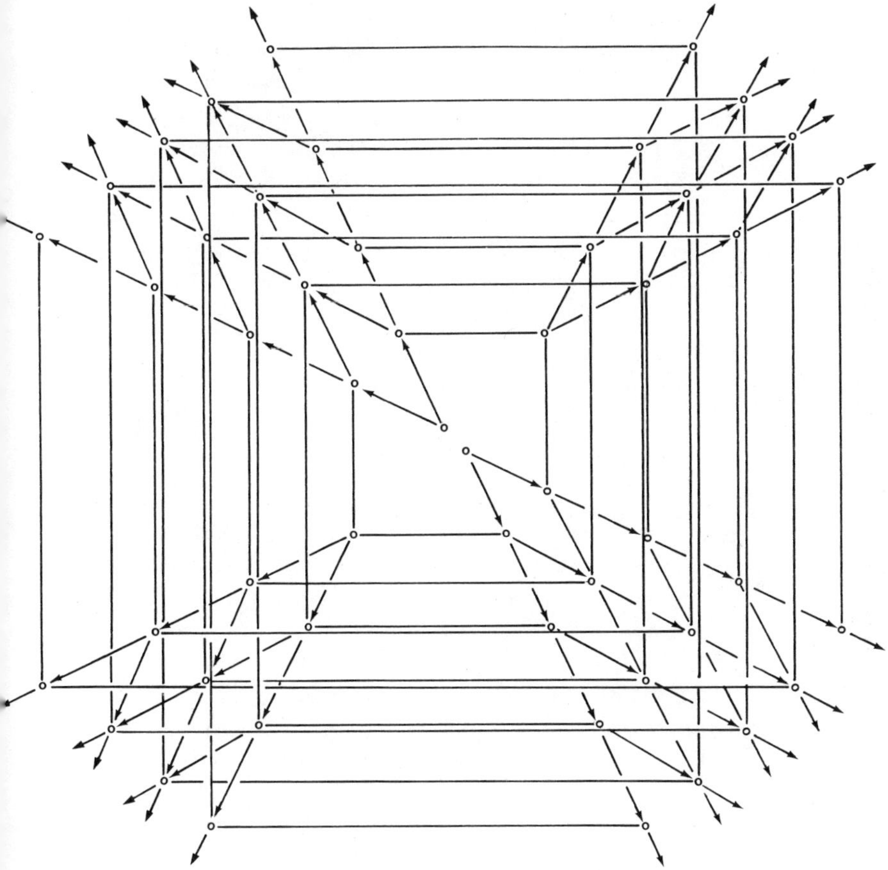

Diagram 2

their duals. Hence pseudo-semilattices form a variety and free pseudo-semilattices exist. Meakin and Pastijn [F] have determined the structure of the free pseudo-semilattice on two generators: this result is contained in the following theorem.

THEOREM 3.4. *Let L denote the free commutative monoid on two generators $\{x,y\}$, so that $L = \{x^m y^n : m,n \geq 0\}$ (where we interpret x^0 and y^0 as 1, the identity of the monoid L). Relative to the natural ordering*

$$x^m y^n \leq x^p y^q \quad iff \quad m \geq p \quad and \quad n \geq q,$$

L becomes a semilattice with identity 1. Let $I = \Lambda = \{1,2\}$ and for each $(i,\lambda) \in I \times \Lambda$ let $E_{i\lambda} = \{(x^m y^n)_{i\lambda} : m,n \in \mathbb{N}\}$ be a semilattice isomorphic by the mapping $(x^m y^n)_{i\lambda} \mapsto x^m y^n$ to L: again, for $(i,\lambda) \in I \times \Lambda$ let $L_\lambda = R_i = L$. Let F_2 be the pseudo-semilattice $F_2 = E(L_\lambda;R_i;E_{i\lambda};\phi_{i\lambda};\psi_{i\lambda};I,\Lambda)$ where the ϕ and ψ mappings are defined as follows:

$$(x^m y^n)_{11}\phi_{11} = (x^m y^n)_{11}\psi_{11} = (x^m y^n)_{22}\phi_{22} = (x^m y^n)_{22}\psi_{22} = x^m y^n,$$

$$(x^m y^n)_{12}\phi_{12} = x^{m+1} y^n, \qquad (x^m y^n)_{12}\psi_{12} = x^m y^{n+1},$$

$$(x^m y^n)_{21}\phi_{21} = x^{m+1} y^n, \qquad (x^m y^n)_{21}\psi_{21} = x^m y^{n+1}.$$

Then F_2 is biorder isomorphic to the free pseudo-semilattice on two generators.

The pseudo-semilattice F_2 is pictured in diagram 2. We close this section by stating the following theorem, which can easily be proved by using a slightly modified version of construction 3.3.

THEOREM 3.5. *Let E be a pseudo-semilattice on n generators. Then there is a pseudo semilattice $\overline{E}$ on n generators such that $\overline{E}$ is an image of F_n (the free pseudo-semilattice on n generators) by semilattices and E is an image of $\overline{E}$ by rectangular biordered sets.*

4. Coextension of biordered sets by rectangular biordered sets

We turn now to the question fo constructing all coextensions of a given biordered set E by rectangular biordered sets. A general construction of such coextensions, together with a construction of all coextensions of a regular semigroup by rectangular bands, has been given by Meakin and Nambooripad, in [B] and [C]. The general case is quite difficult and technical and we shall only consider two special cases of this construction here.

We first recall that Clifford [1976] has defined a *solid biordered set* E to be one in which $R \circ L = L \circ R$ and Yamada [1979a] has defined a regular semigroup S to be *quasi-orthodox* if S is a coextension of an inverse semigroup by completely simple semigroups (i.e. if there is a homomorphism ϕ from S onto an inverse semigroup T such that for each $e \in E(S)$, $e(\phi \circ \phi^{-1})$ is a completely simple subsemigroup of S: such homomorphisms were called "strictly compatible" by Nambooripad [G] who has provided an elegant alternative characterization of them. From Clifford [1976] and Hall's supplement to Yamada's paper [1979] and Hall [1973a] we have the following result.

THEOREM 4.1. *The following are equivalent for a biordered set E:*

(a) E is a solid biordered set;

(b) $E = E(S)$ for some completely regular semigroup S;

(c) every regular semigroup S with $E(S) = E$ is quasi-orthodox;

(d) E is a coextension of a semilattice by rectangular biordered sets.

A construction of solid biordered sets is provided in the following theorem due to Meakin and Nambooripad [B].

THEOREM 4.2. *Let F be a semilattice and for each $\alpha \in F$ let $E_\alpha = I_\alpha \times \Lambda_\alpha$ be a rectangular biordered set such that $I_\alpha \cap I_\beta = \Box$ and $\Lambda_\alpha \cap \Lambda_\beta = \Box$ if $\alpha \neq \beta$. Suppose that for each $\alpha, \beta \in F$ with $\beta \leqslant \alpha$ and each $e \in E_\alpha$ there are mappings $\phi^e_\beta : I_\beta \to I_\beta$ and $\psi^e_\beta : \Lambda_\beta \to \Lambda_\beta$ which satisfy the following condition (S1) and its dual:*

(S1) If $e = (i,\lambda) \in E_\alpha$ then ϕ^e_α is a constant map with value i.

Let $E = \bigcup_{\alpha\in F} E_\alpha$, $e = (i,\lambda) \in E_\alpha$ and $f = (j,\mu) \in E_\beta$. Define $f\ \omega^r\ e$ iff $\phi^e_\beta(j) = j$ and $f\ \omega^\ell\ e$ iff $\mu\psi^e_\beta = \mu$. If $f\,\omega^r\,e$ define $ef = f = (j,\mu)$ and $fe = (j,\mu\psi^e_\beta)$. If $f\ \omega^\ell\ e$ define $fe = f = (j,\mu)$ and $ef = (\phi^e_\beta(j),\mu)$. Then the products ef, fe defined when $f\ \kappa\ e$ or $e\ \kappa\ f$ are well-defined.

Suppose, in addition, that the ϕ and ψ mappings satisfy the *following conditions (S2) and (S3) and their duals:*

(S2) *If* $e \in E_\alpha$, $f \in E_\beta$, $f \; \omega^r \; e$ *and* $\gamma \leqslant \beta$ *then*

$$\phi^e_\gamma \, \phi^f_\gamma = \phi^f_\gamma \quad \text{and} \quad \psi^e_\gamma \, \psi^f_\gamma = \psi^f_\gamma .$$

(S3) *If* $e \in E_\alpha$, $f \in E_\beta$, $\gamma = \alpha\beta$, $\delta \leqslant \gamma$ *and* $h \in E_\gamma \cap M(e,f)$, *then*

$$\phi^e_\delta \, \phi^f_\delta = \phi^{eh}_\delta \, \phi^{hf}_\delta \text{ and } \psi^e_\delta \, \psi^f_\delta = \psi^{eh}_\delta \, \psi^{hf}_\delta .$$

Then, relative to the basic products defined above, E is a solid biordered set (which is a coextension of F by rectangular biordered sets). Every solid biordered set is obtained this way.

REMARK. We need to impose an additional restriction on the ϕ and ψ mappings in order that E admits a band structure (see [B]). This shows clearly the distinction between "solid biordered sets" and "orthodox biordered sets" (i.e. the biordered sets of bands).

We turn now to a brief discussion of a second special case of the general construction of coextensions of biordered sets by rectangular biordered sets.

DEFINITION 4.3. If E and F are biordered sets we say that E is a *normal* coextension of F if there is a bimorphism $\theta \colon E \twoheadrightarrow F$ from E onto F such that the following condition is satisfied:

(N) if $\beta \; \omega \; \alpha$ in F, $e \in E$ and $e\theta = \alpha$ then there is a *unique* element $f \in E$ such that $f\theta = \beta$ and $f \; \omega \; e$.

It is clear that if E is a normal coextension of F (by means of a bimorphism $\theta \colon E \twoheadrightarrow F$) then ω is trivial on each θ-class of E and so E is a coextension of F by rectangular biordered sets. It is well-known that a band E is normal (in the sense of Yamada and Kimura [1958]) iff its biordered set is a normal coextension of a semilattice by rectangular biordered sets. Another pleasant fact about normal coextensions is contained in the following theorem of Meakin and Nambooripad [D].

THEOREM 4.4. *A coextension of a pseudo-semilattice by rectangular biordered sets is again a pseudo-semilattice if and only if the coextension is normal.*

Normal coextensions of arbitrary biordered sets may be constructed by the following theorem of Meakin and Nambooripad [B].

THEOREM 4.5. *Let F be a biordered set and for each $\alpha \in F$ let $E_\alpha = I_\alpha \times \Lambda_\alpha$ be a rectangular biordered set such that $I_\alpha = I_\beta$ if $\alpha\ R\ \beta$; $I_\alpha \cap I_\beta = \square$ if $(\alpha,\beta) \notin R$; $\Lambda_\alpha = \Lambda_\beta$ if $\alpha\ L\ \beta$; and $\Lambda_\alpha \cap \Lambda_\beta = \square$ if $(\alpha,\beta) \notin L$. Let $\chi = \{\chi_{\alpha,\beta}\colon E_\alpha \to E_\beta\colon \beta\ \omega\ \alpha\}$ be a family of mappings $\chi_{\alpha,\beta}\colon E_\alpha \to E_\beta$ defined if $\beta\ \omega\ \alpha$ in F and assume that χ satisfies the following conditions and their duals:*

(N1) $\chi_{\alpha,\alpha}$ is the identity on E_α for all $\alpha \in F$;

(N2) let $e \in E_\alpha$, $f \in E_\beta$, $\alpha\ R\ \beta$, $\gamma\ \omega\ \alpha$, $\delta\ \omega\ \beta$ and $\gamma\ R\ \delta$: then $e\chi_{\alpha,\gamma}\ R\ f\chi_{\beta,\delta}$ in $E_\gamma \cup E_\delta$ if $e\ R\ f$ in $E_\alpha \cup E_\beta$;

(N3) if $\gamma\ \omega\ \beta\ \omega\ \alpha$ then $\chi_{\alpha,\beta}\ \chi_{\beta,\gamma} = \chi_{\alpha,\gamma}$.

On $E = \bigcup_{\alpha\in F} E_\alpha$ define quasi-orders ω^r and ω^ℓ as follows: if $e \in E_\alpha$ and $f \in E_\beta$ define

$$f\ \omega^r\ e \text{ iff } \beta\ \omega^r\ \alpha \text{ and } e\chi_{\alpha,\beta\alpha}\ R\ f \text{ in } E_\beta \cup E_{\beta\alpha},$$

$$f\ \omega^\ell\ e \text{ iff } \beta\ \omega^\ell\ \alpha \text{ and } e\chi_{\alpha,\alpha\beta}\ L\ f \text{ in } E_\beta \cup E_{\alpha\beta}.$$

For $f\ \omega^r\ e$ define $ef = f$ and $fe = e\chi_{\alpha,\beta\alpha}$: for $f\ \omega^\ell\ e$ define $ef = e\chi_{\alpha,\alpha\beta}$ and $fe = f$. Relative to these products, E is a biordered set which is a normal coextension of F (by the rectangular biordered sets E_α $(\alpha \in F)$). Every normal coextension of F is obtained this way.

A coextension of a pseudo-inverse semigroup by rectangular bands has been called "locally orthodox" by Meakin and Nambooripad [D] since such a semigroup has the property that $\omega^r(e)$ and $\omega^\ell(e)$ are bands for each idempotent e. A locally orthodox semigroup is pseudo-inverse if and only if the coextension is normal. Thus, in general, the process of taking coextensions of pseudo-semilattices by rectangular biordered sets yields biordered sets which are not pseudo-semilattices. It is at present not known whether the process of taking successive coextensions or images of biordered sets by rectangular biordered sets or semilattices terminates (in the sense that no new biordered sets are obtained) in a finite number of steps.

5. Finite biordered sets

We close the paper with an observation due to McAlister and Meakin (unpublished) that finite biordered sets may in principle be built from rectangular biordered sets and semilattices.

For each class $\mathcal{E}$ of biordered sets we let $\overline{\mathcal{E}}$ be the smallest class of biordered sets such that

(1) $\mathcal{E} \subseteq \overline{\mathcal{E}}$;

(2) if $E \in \overline{\mathcal{E}}$ then $E \cup \{1\} \in \overline{\mathcal{E}}$ (here $E \cup \{1\}$ is the biordered set obtained from E by adjoining an identity 1 in the obvious way);

(3) if $E \in \overline{\mathcal{E}}$ then every image of E by rectangular biordered sets is in $\overline{\mathcal{E}}$;

(4) if E is a rectangular biordered set then every coextension of E by biordered sets in $\overline{\mathcal{E}}$ is in $\overline{\mathcal{E}}$ (i.e. if F is a biordered set admitting a bimorphism θ: $F \twoheadrightarrow E$ from F onto E such that each θ-class of F is a biordered set in $\overline{\mathcal{E}}$, then $F \in \overline{\mathcal{E}}$).

Denote the trivial biordered set by e: the observation of Meakin and McAlister is the following.

THEOREM 5.1. *Every finite biordered set is in* $\overline{\{e\}}$.

Proof. Let E be a finite biordered set. If E has an identity 1 then $E\setminus\{1\}$ is a biordered set of smaller cardinality than E and E is obtained from $E\setminus\{1\}$ by adjoining an identity. So suppose that E does not have an identity. Form the biordered set $\overline{E}$ of construction 3.3: $\overline{E}$ is a coextension of E by rectangular biordered sets. In addition, $\overline{E}$ is a coextension of the rectangular biordered set $E \times E$ by biordered sets of the form $M(e,f)$ for $e,f \in E$. Since E does not have an identity, each biordered set $M(e,f)$, for $e,f \in E$, has cardinality less than the cardinality of E. It follows by induction that $E \in \overline{\{e\}}$.

Theorem 5.1 suggests that it may be fruitful to develop some kind of a "complexity" theory for (at least finite) biordered sets: the complexity of a biordered set should be a measure of the number of steps required to build the biordered set from rectangular biordered sets and semilattices by means of the processes discussed in this paper (or similar processes). For example, rectangular biordered sets and semilattices should be assigned complexity 0, solid biordered sets should have complexity ≤1, pseudo-semilattices should have complexity ≤2, biordered sets of locally orthodox semigroups should have complexity ≤3, etc. No work along these lines has yet been done, as far as this author is aware. Of course the kind of complexity theory suggested here is rather different from the classical Krohn-Rhodes-Tilson complexity theories for finite semigroups (see, for example, Tilson [1971]) since no process analogous to the process of "division" is envisaged.

References

1958 M. Yamada and N. Kimura, Note on idempotent semigroups II, *Proc. Japan. Acad.* 34 (1958), 110-112.

1967 M. Yamada, Regular semigroups whose idempotents satisfy permutation identities, *Pacific J. Math.* 21 (1967), 371-392.

1970 W.D. Munn, Fundamental inverse semigroups, *Quart. J. Math.* (Oxford) (2) 21 (1970), 157-70.

1971 B. Tilson, Decomposition and complexity of finite semigroups, *Semigroup Forum* 3 (1971), 189-250.

1972 B.M. Schein, Pseudo-semilattices and pseudo-lattices, *Izv. Vysš. Učebn. Zaved Mathematika* 2 (117) (1972), 81-94 (in Russian).

1973a T.E. Hall, On regular semigroups, *J. Algebra* 24 (1973), 1-24.

1973b Y. Zalcstein, Locally testable semigroups, *Semigroup Forum* 5 (1973), 216-227.

1976 A.H. Clifford, The fundamental representation of a completely regular semigroup, *Semigroup Forum* 12 (1976), 341-346.

1978 K. Byleen, J. Meakin and F. Pastijn, The fundamental four-spiral semigroup, *J. Algebra* 54 (1978), 6-26.

1979a M. Yamada, On a certain class of regular semigroups, *Proc. Conference on regular semigroups*, Northern Illinois University, De Kalb (1979).

1979b K.S.S. Nambooripad, Structure of regular semigroups I, *Memoirs Amer. Math. Soc.* 244 (1979).

TO BE PUBLISHED

[A] K. Byleen, J. Meakin and F. Pastijn, Building bisimple idempotent-generated semigroups, *J. Algebra* (to appear).

[B] J. Meakin and K.S.S. Nambooripad, Coextensions of regular semigroups by rectangular bands I, submitted for publication.

[C] J. Meakin and K.S.S. Nambooripad, Coextensions of regular semigroups by rectangular bands II, to be submitted for publication.

[D] J. Meakin and K.S.S. Nambooripad, Coextensions of pseudo-inverse semigroups by rectangular bands, submitted for publication.

[E] J. Meakin and F. Pastijn, The structure of pseudo-semilattices, submitted for publication.

[F] J. Meakin and F. Pastijn, The free pseudo-semilattice on two generators, in preparation.

[G] K.S.S. Nambooripad, The natural partial order on a regular semigroup, submitted for publication.

[H] K.S.S. Nambooripad, Pseudo-semilattices and biordered sets, submitted for publication.

[I] F. Pastijn, Rectangular bands of inverse semigroups, submitted for publication.

SEMIGROUPS WHOSE CONGRUENCES FORM A CHAIN AND WHICH ARE EXTENSIONS OF CONGRUENCE–FREE SEMIGROUPS

Takayuki Tamura

Department of Mathematics, University of California, Davis, California, U.S.A., 95616

Abstract

This paper studies finite semigroups which are ideal extensions of congruence-free semigroups and which are such that their set of congruences forms a chain, and shows how to construct all such semigroups.

1. Introduction

A semigroup is called a Δ-semigroup if its congruences form a chain under inclusion. There are many well-known examples of finite Δ-semigroups, for example, symmetric groups, symmetric inverse semigroups, full transformation semigroups, each on a finite set. The study of Δ-semigroups was initiated in the commutative case by Schein [1969a] and Tamura [1969b] independently. These results were extended to the medial case (namely, to semigroups satisfying the identity $xyzu = xzyu$) by Etterbeek [1970] and to the exponential case (satisfying identity $(xy)^n = x^n y^n$ for all n) by Trotter [1976]. Inverse Δ-semigroups with principal series were studied by Trotter and the author [1977]. Quite recently Bonzini and Spoletini [1979a] have studied E-m Δ-semigroups (namely, those satisfying the identity $(xy)^m = x^m y^m$). Part of our results on finite Δ-semigroups was reported in [Tamura 1979b].

In this paper we study two special cases of Δ-semigroups: (i) completely [0-] simple Δ-semigroups and (ii) finite Δ-semigroups T which are ideal extensions of congruence-free semigroups S by 0-simple semigroups V. Completely simple Δ-semigroups are only the trivial cases, that is, Δ-groups, and right or left zero semigroups of order 2. Completely 0-simple Δ-semigroups are characterized by their sandwich

ISBN 0-12-319450-4

matrices. Since property Δ is preserved under homomorphisms, V must be a Δ-semigroup. But Δ is not in general preserved by ideals. Ideal extensions of a completely 0-simple semigroup by a completely 0-simple semigroup were studied by Warne [1966] and Petrich [1967b]. In this paper, however, the author intends to construct directly and concretely finite Δ-semigroups T, which are ideal extensions of congruence-free semigroups S by 0-simple semigroups V, not as extensions of a given S but as pre-images of a given V. The case where S is congruence-free is the first step toward the study of the general case.

2. Preliminaries

In this section we give some basic results on Δ-semigroups.

LEMMA 2.1 (T. Tamura [1969b]). *A homomorphic image of a Δ-semigroup is a Δ-semigroup.*

Let S be a proper ideal of T. For each $a \in T$ define $\phi_a(\psi_a) : S \to S$ by $x\phi_a = xa$ ($\psi_a x = ax$). Then $\phi_a(\psi_a)$ is a right (left) translation of S. If $\phi_a \neq \phi_b$ for some $a, b \in S$, we say the right regular representation of S is not trivial.

When ρ and σ are congruences, $a \; \rho\setminus\sigma \; b$ means $a \; \rho \; b$ but $a \not\sigma b$. Write $\omega_D = D \times D$, $\iota_D = \{(x,x) : x \in D\}$.

PROPOSITION 2.2. *Let S be a proper ideal of T such that the right (left) regular representation of S is not trivial. If T is a Δ-semigroup, then*

(2.2.1) *for all $p \in T\setminus S$, $\phi_p(\psi_p)$ is not inner and*

(2.2.2) *the mapping $p \mapsto \phi_p$ ($p \mapsto \psi_p$) is one-to-one for $p \in T\setminus S$.*

Proof. Necessarily $|S| > 1$. By assumption $\phi_a \neq \phi_b$ for some $a, b \in S$. Suppose ϕ_p is inner for some $p \in T\setminus S$, say $\phi_p = \phi_c$ for some $c \in S$. Then we have two incomparable congruences ρ and σ on T as follows: Let σ be the Rees congruence on T modulo S. Define ρ by $x \; \rho \; y$ if and only if $\phi_x = \phi_y$. Then $c \; \rho\setminus\sigma \; p$ and $a \; \sigma\setminus\rho \; b$, a contradiction. Next suppose $\phi_p = \phi_q$ for some $p, q \in T\setminus S$, $p \neq q$. Then $p \; \rho\setminus\sigma \; q$ and $a \; \sigma\setminus\rho \; b$, again a contradiction.

Let T be a set of transformations of a set X. An equivalence ξ on X is called T-compatible if $x \,\xi\, y$ implies $xf \,\xi\, yf$ for all $f \in T$ and hence for all f of the semigroup $\bar{T}$ generated by T.

Let S be an ideal of a semigroup T, and $V = T/S$, $V^* = V\setminus\{0\}$. Let $\Phi = \{\phi_a : a \in V^*\}$, $\Psi = \{\psi_a : a \in V^*\}$.

PROPOSITION 2.3. *Let T be an ideal extension of S by V. Then T is a Δ-semigroup if and only if the following hold.*

(2.3.1) *V is a Δ-semigroup.*

(2.3.2) *All Φ- and Ψ-compatible congruences ρ on S form a chain.*

(2.3.3) *If $a \in T$ and $b \in T\setminus S$, with $a \neq b$, then the congruence ζ_0 on T generated by (a,b) contains ω_S.*

Proof. Necessity. Condition (2.3.1) follows from lemma 2.1. Condition (2.3.2) follows from the fact that $\rho \to \rho \cup \iota_T$ is one-to-one and order-preserving. To show condition (2.3.3) we let $\rho = \zeta_0|S$, and suppose $\rho \neq \omega_S$. Then ζ_0 is incomparable with the Rees-congruence on T modulo S. Hence $\rho = \omega_S$.

Sufficiency. By (2.3.3), every congruence ζ on T satisfies either $\zeta \supset \omega_S$ or $\zeta = \rho \cup \iota_T$ where $\rho = \zeta|S \subset \omega_S$. All ζ's containing ω_S form a chain by (2.3.1); all the others form a chain by (2.3.2). It follows that T is a Δ-semigroup.

A semigroup D is called *congruence-free* if it has no congruences except ι_D and ω_D.

COROLLARY 2.4. *In proposition 2.3 if V is 0-simple then condition (2.3.3) can be omitted. If S is congruence-free and V is 0-simple then (2.3.2) and (2.3.3) can be omitted.*

LEMMA 2.5. *If S is a Δ-semigroup then S is subdirectly irreducible.*

Proof. If S is subdirectly reducible then the two congruences induced by projections are incomparable.

We now consider a Δ-semigroup and its semilattice decomposition.

PROPOSITION 2.6. *If S is a Δ-semigroup, then S is either a semilattice-indecomposable semigroup or the semilattice of two semilattice-indecomposable semigroups.*

This follows from the fact that a Δ-semilattice is of order at most 2.

3. Completely [0–] simple semigroups

A completely simple semigroup S is regarded as a Rees matrix semigroup [Clifford and Preston 1961; Petrich 1973a]:

$$S = M(I,G,M:P) = \{(i,x,j) : i \in I, x \in G, j \in M\}$$

where G is called the basic group and P is the so-called sandwich $M \times I$ -matrix, $P = (p_{ji})$, $j \in M$, $i \in I$, $p_{ji} \in G$. The operation in S is given by $(i,x,j)(k,y,l) = (i,xp_{jk}y,l)$. We can assume P is normalized, and each semigroup is not trivial.

THEOREM 3.1. *Let S be a completely simple semigroup. Then S is a Δ-semigroup if and only if S is isomorphic to one of the following:*

(3.1.1) *a Δ-group,*

(3.1.2) *a right zero semigroup of order 2,*

(3.1.3) *a left zero semigroup of order 2.*

Proof. Assume S is a Δ-semigroup. By lemma 2.1, the rectangular band $I \times M$ is a Δ-semigroup. However, by lemma 2.5, either $|I| = 1$ or $|M| = 1$. Assume $|I| = 1$. It follows that S is isomorphic to a right group $G \times M$. By lemma 2.5 $|M| = 1$ or $|G| = 1$. The conclusion immediately follows. The converse is obvious.

Completely 0-simple Δ-semigroups without zero divisors can be dealt with as a corollary.

COROLLARY 3.2. *A semigroup S is a completely 0-simple Δ-semigroup without zero divisors if and only if S is isomorphic to one of the following:*

(3.2.1) *a Δ-group with 0 adjoined,*

(3.2.2) *a right zero semigroup of order 2 with 0 adjoined,*

(3.2.3) *a left zero semigroup of order 2 with 0 adjoined.*

Completely 0-Simple Δ-Semigroups. Let $S = M^0(I,G,M;P)$ and let (i,x,j) and (k,y,l) be any elements of S, $i \in I$, $x \in G$, $j \in M$, $k \in I$, $y \in G, l \in M$, with

$$(i,x,j)(k,y,l) = \begin{cases} (i,xp_{jk}y,l), & \text{if } p_{jk} \neq 0, \\ 0, & \text{if } p_{jk} = 0, \end{cases}$$

where 0 is the zero of S. Let γ be the non-universal semilattice congruence on $G^0 = G \cup \{0\}$. Define equivalences ρ_0 and σ_0 on M and I, respectively, as follows:

$j \; \rho_0 \; l$ if and only if $p_{jk} \; \gamma \; p_{lk}$ for all $k \in I$,

$i \; \sigma_0 \; k$ if and only if $p_{ji} \; \gamma \; p_{jk}$ for all $j \in M$.

We assume $P = (p_{ji})$ is normalized in the following sense [Tamura 1960]. There are $\alpha : I \to M$ and $\beta : M \to I$ such that

(1) $\sigma_0 \subseteq \alpha \circ \alpha^{-1}$ and $\rho_0 \subseteq \beta \circ \beta^{-1}$,

(2) $p_{i\alpha, i} = e$ for all $i \in I$,

(3) $p_{j, j\beta} = e$ for all $j \in M$,

where e is the identity element of G.

Let $S = M^0(I,G,M;P)$ where P is normalized. All non-universal congruences on S can be obtained in the following manner [Tamura 1960 or Lallement 1974 or Johnston 1978; also see Clifford and Preston 1967a].

Let ρ and σ be equivalences on M and I respectively such that $\rho \subseteq \rho_0$ and $\sigma \subseteq \sigma_0$. Let τ be a congruence on G such that

$$\{(p_{jk}, p_{lk}) : p_{jk} \neq 0, \; j \; \rho \; l\} \cup \{(p_{ji}, p_{jk}) : p_{ji} \neq 0, \; i \; \sigma \; k\} \subseteq \tau.$$

Given ρ, σ and τ, define η on S as follows:

$0 \; \eta \; 0$

$(i,x,j) \; \eta \; (k,y,l)$ if and only if $i \; \sigma \; k$, $x \; \tau \; y$, $j \; \rho \; l$.

Then η is a congruence on S and every non-universal congruence on S can be obtained in this manner. Thus η is denoted by

$$\eta = [\sigma, \tau, \rho].$$

Obviously $[\sigma_1, \tau_1, \rho_1] \subseteq [\sigma_2, \tau_2, \rho_2]$ if and only if $\sigma_1 \subseteq \sigma_2$, $\tau_1 \subseteq \tau_2$ and $\rho_1 \subseteq \rho_2$.

THEOREM 3.3. *Let S be a completely 0-simple semigroup with zero divisors and with normalized P : $S = M^0(I,G,M;P)$. Then S is a Δ-semigroup if and only if G is a Δ-group and one of the following holds:*

(3.3.1) $\rho_0 = \iota_M$, $\sigma_0 = \{(i_1,i_2),(i_2,i_1)\} \cup \iota_I$ *for some* $i_1, i_2 \in I$, $i_1 \neq i_2$, *and* $\{(p_{ji_1}, p_{ji_2}) : j \in M$ *with* $p_{ji_1} \neq 0\}$ *generates* ω_G.

(3.3.2) $\sigma_0 = \iota_I$, $\rho_0 = \{(j_1,j_2),\ (j_2,j_1)\} \cup \iota_M$ *for some* $j_1,\ j_2 \in M$, $j_1 \neq j_2$, *and* $\{(p_{j_1 i}, p_{j_2 i}) : i \in I \text{ with } p_{j_1 i} \neq 0\}$ *generates* ω_G.

(3.3.3) $\sigma_0 = \iota_I$ *and* $\rho_0 = \iota_M$.

Proof. Necessity. If τ is a congruence on G, then $[\iota_I,\tau,\iota_M]$ is a congruence on S. It follows that G is a Δ-group. To make $[\iota_I,\omega_G,\rho_0]$ comparable with $[\sigma_0,\omega_G,\iota_M]$ it must hold that $\rho_0 = \iota_M$ or $\sigma_0 = \iota_I$. Assume $\rho_0 = \iota_M$ but $\sigma_0 \neq \iota_I$. Suppose $i_1\sigma_0 i_2$, $i_1 \neq i_2$, and $i_3\sigma_0 i_4$, $i_3 \neq i_4$ where at least one of i_3 and i_4 is distinct from i_1 and i_2. Define σ_1 and σ_2 by

$$\sigma_1 = \{(i_1,i_2),\ (i_2,i_1)\} \cup \iota_I,\quad \sigma_2 = \{(i_3,i_4),(i_4,i_3)\} \cup \iota_I.$$

Then $i_1\ \sigma_1\setminus\sigma_2\ i_2$ and $i_3\ \sigma_2\setminus\sigma_1\ i_4$, hence $[\sigma_1,\omega_G,\iota_M]$ and $[\sigma_2,\omega_G,\iota_M]$ are incomparable. This contradicts the assumption. Therefore if $\sigma_0 \neq \iota_I$ there is only one non-singleton σ_0-class and it consists of exactly two elements, say, i_1, i_2, $i_1 \neq i_2$. Suppose $\{(p_{ji_1}, p_{ji_2}) : j \in M,\ p_{ji_1} \neq 0\}$ generates a congruence τ_1 on G such that $\tau_1 \neq \omega_G$. Then $[\sigma_0,\tau_1,\iota_M]$ and $[\iota_I,\omega_G,\iota_M]$ are incomparable, a contradiction. Therefore $\tau_1 = \omega_G$.

Sufficiency. Assume (3.3.1). All congruences on S which are not universal are obtained as follows:

either $\eta_\omega = [\sigma_0,\omega_G,\iota_M]$ where $\sigma_0 = \{(i_1,i_2),(i_2,i_1)\} \cup \iota_I$, $i_1 \neq i_2$,

or $\eta_\tau = [\iota_I,\tau,\iota_M]$ where τ is any congruence on G.

Since $\tau_1 \subseteq \tau_2$ implies $[\iota_I,\tau_1,\iota_M] \subseteq [\iota_I,\tau_2,\iota_M]$, and moreover $[\iota_I,\tau,\iota_M] \subset [\sigma_0,\omega_G,\iota_M]$ for all τ, therefore since G is a Δ-group, S is a Δ-semigroup. The dual result is obtained from (3.3.2). In case (3.3.3) all congruences ($\neq \omega_S$) can be obtained as η_τ, so S is a Δ-semigroup. This completes the proof.

REMARK 3.4. In theorem 3.3 normalization of P is essential. For example let $|G| = 2$, $G = \{1,\alpha\}, \alpha^2 = 1$.

$M^0(3,G,3;\ \begin{pmatrix}1&1&0\\1&\alpha&1\\0&0&1\end{pmatrix})$ is a Δ-semigroup.

$M^0(3,G,3;\ \begin{pmatrix}1&\alpha&0\\1&\alpha&1\\0&0&1\end{pmatrix})$ is not a Δ-semigroup although (3.3.1) is fulfilled.

$M^0(3,H,3;\ \begin{pmatrix}1&1&0\\1&1&1\\0&0&1\end{pmatrix})$ is a Δ-semigroup if and only if $H = \{1\}$.

$M^0(I,G,M;P)$ is congruence-free if and only if $\sigma_0 = \iota_I$, $\rho_0 = \iota_M$ and $G = \{e\}$ [Tamura 1956].

4. 0– minimal ideals

In this section we find a necessary condition for a completely 0-simple semigroup to be a 0-minimal ideal of a Δ-semigroup.

First we review bitranslations of $S = M^0(I,G,M;P)$ [Petrich 1973a]. Let $M^0 = M \cup \{0\}$, $I^0 = I \cup \{0\}$, $G^0 = G \cup \{0\}$. Let $g : M^0 \to G^0$, $r : I^0 \to G^0$ be mappings of M^0 and I^0 into G^0, respectively, satisfying $0g = 0$, $r0 = 0$. Let $f : M^0 \to M^0$ and $q : I^0 \to I^0$ be transformations of M^0 and I^0, respectively, satisfying $0f = 0$ and $q0 = 0$. Moreover assume

$$jf = 0 \text{ if and only if } jg = 0,$$

$$qi = 0 \text{ if and only if } ri = 0.$$

Given g, f, r, q, define $\phi : S \to S$ and $\psi : S \to S$ as follows:

$$(i,x,j)\phi = (i,x(jg),jf)$$

$$\psi(i,x,j) = (qi,(ri)x,j).$$

Then $\phi(\psi)$ is a right (left) translation of S and every right (left) translation of S can be obtained in this manner; ϕ and ψ are denoted by

$$\psi = (q,r],\ \phi = [g,f).$$

LEMMA 4.1 (M. Petrich [1973a]). *In $M^0(I,G,M;P)$, $(q,r]$ is linked with $[g,f)$ if and only if*

(4.1.1) $(jg)\, p_{jf,k} = p_{j,qk}(rk)$ *for all $j \in M$, $k \in I$.*

Note ϕ and ψ are respectively associated with an $M \times M$-matrix $A(\phi)$ and $I \times I$-matrix $B(\psi)$:

$$A(\phi) = (a_{ji}),\ j,\ i \in M, \text{ defined by}$$

$$a_{ji} = \begin{cases} jg & \text{if } i = jf \\ 0 & \text{if } i \neq jf, \end{cases}$$

$B(\psi) = (b_{ik})$, $i, k \in I$, defined by

$$b_{ik} = \begin{cases} rk \text{ if } i = qk \\ 0 \text{ if } i \neq qk. \end{cases}$$

Condition (4.1.1) can be restated as:

(4.1.2) $A(\phi) \cdot P = P \cdot B(\psi)$.

Denote by Φ the set of all right translations of $M^0(I,G,M;P)$ and by Ψ the set of all its left translations.

Let $F = \{f : [g,f) \in \Phi \text{ for some } g\}$, $G_\Phi = \{g : [g,f) \in \Phi \text{ for some } f\}$,

$Q = \{q : (q,r] \in \Psi \text{ for some } r\}$, $G_\Psi = \{r : (q,r] \in \Psi \text{ for some } q\}$.

LEMMA 4.2. *Under normalization of P, $[\sigma,\tau,\rho]$ is Φ- and Ψ-compatible if and only if*

ρ is F-compatible,

σ is Q-compatible,

$j_1 \; \rho \; j_2$ implies $j_1 g \; \tau \; j_2 g$ for all $g \in G_\Phi$,

$i_1 \; \sigma \; i_2$ implies $ri_1 \; \tau \; ri_2$ for all $r \in G_\Psi$.

The following gives a necessary condition for S to be a 0-minimal ideal of a Δ-semigroup.

THEOREM 4.3. *Let T be a finite semigroup with zero and S a 0-minimal ideal of T, with $S \neq T$. If T is a Δ-semigroup, then S is either a null semigroup or a completely 0-simple semigroup $M^0(I,G,M;P)$ with the following properties:*

(4.3.1) *S has zero divisors,*

(4.3.2) *G is a Δ-group,*

(4.3.3) *$\rho_0 = \iota_M$ or $\sigma_0 = \iota_I$ or both.*

Proof. Assume that T is a Δ-semigroup and S is not a null semigroup. Let $S = M^0(I,G,M;P)$. Let us consider when a congruence η on S can be extended to a congruence $\bar{\eta}$ on T in the sense that $\bar{\eta} = \eta \cup \iota_T$. If τ is any congruence on G, $\eta_\tau = [\iota_I,\tau,\iota_M]$ is a congruence on S. Let ϕ and ψ be right and left translations on S, respectively. It is easy to see that if $(i,x,j) \; \eta_\tau \; (i,y,j)$ then $(i,x,j)\phi \; \eta_\tau \; (i,y,j)\phi$ and $\psi(i,x,j) \; \eta_\tau \; \psi(i,y,j)$. It follows that

η_τ can be extended to $\bar{\eta}_\tau$ on T. Consequently G must be a Δ-group. Consider two congruences η_1 and η_2 on S:

$$\eta_1 = [\sigma_0,\omega_G,\iota_M],\ \eta_2 = [\iota_I,\omega_G,\rho_0].$$

Let $\phi = [g,f)$, $\psi = (q,r]$ and assume that ϕ and ψ are linked. We want to show that η_1 and η_2 can be extended to $\bar{\eta}_1$ and $\bar{\eta}_2$ on T, respectively. If $(i_1,x,j)\ \eta_1\ (i_2,y,j)$, so that $i_1\ \sigma_0\ i_2$, then

$$(i_1,x,j)\phi = (i_1,x(jg),jf)\ \eta_1\ (i_2,y(jg),jf) = (i_2,y,j)\phi, \text{ and}$$

thus η_1 is compatible under ϕ. Next, under ψ,

$$\psi(i_1,x,j) = (qi_1,(ri_1)x,j),$$

$$\psi(i_2,y,j) = (qi_2,(ri_2)y,j).$$

We assume that ψ and ϕ are linked. If the following conditions (i) and (ii) are proved, then, using lemma 4.2, we see that η_1 will be also compatible under ψ.

(i) If $i_1\ \sigma_0\ i_2$ then $ri_1\ \gamma\ ri_2$ where γ was defined in section 3.

(ii) If $i_1\ \sigma_0\ i_2$ and if $ri_1 \neq 0$, then $qi_1\ \sigma_0\ qi_2$.

Suppose that $i_1\ \sigma_0\ i_2$ but $ri_1 \not\gamma\ ri_2$, say $ri_1 = 0$ but $ri_2 \neq 0$. There is $j_1 \in M$ such that $p_{j_1,qi_2} \neq 0$; then $p_{j_1,qi_2}(ri_2) \neq 0$, and, by (4.1.1),

$$(j_1g)p_{j_1f,i_1} = p_{j_1,qi_1}(ri_1) = 0$$

and $$(j_1g)p_{j_1f,i_2} = p_{j_1,qi_2}(ri_2) \neq 0.$$

But, by definition of σ_0, $p_{j_1f,i_1}\ \gamma\ p_{j_1f,i_2}$, hence $(j_1g)p_{j_1f,i_1}\ \gamma\ (j_1g)p_{j_1f,i_2}$. By the above two equalities, we arrive at a contradiction. Thus we have proved (i). Assume $i_1\ \sigma_0\ i_2$ and $ri_1 \neq 0$. Then $ri_2 \neq 0$ by (i). From (4.1.1) and definition of σ_0, we have the following:

For all $j \in M$, $(jg)p_{jf,i_1}\ \gamma\ (jg)p_{jf,i_2}$ implies $p_{j,qi_1}(ri_1)\ \gamma\ p_{j,qi_2}(ri_2)$.

Since $(ri_1)^{-1}\ \gamma\ (ri_2)^{-1}$, $p_{j,qi_1}\ \gamma\ p_{j,qi_2}$ for all $j \in M$, hence $qi_1\ \sigma_0\ qi_2$. Thus we have proved (ii). Accordingly $\psi(i_1,x,j)\ \eta_1\ \psi(i_2,y,j)$, so η_1 can be extended to $\bar{\eta}_1$ on T. By the dual argument, we can show η_2 can be extended to $\bar{\eta}_2$ on T. However,

$\bar{\eta}_1$ and $\bar{\eta}_2$ are incomparable. Therefore if T is a Δ-semigroup, we must have $\rho_0 = \iota_M$ or $\sigma_0 = \iota_I$. Suppose S contains no zero divisors; S has a congruence ξ such that S/ξ has an identity element (in fact, is a semilattice of order 2). Then ξ can be extended to $\bar{\xi}$ on T, and $\bar{\xi}$ is incomparable with the Rees-congruence modulo S. Therefore S contains zero divisors.

5. Ideal extensions of congruence–free semigroups

Throughout this section all semigroups are finite, whence 0-simple semigroups are necessarily completely 0-simple. Let V be a finite 0-simple Δ-semigroup. Our claim is this: for V, construct a transformation semigroup S over a finite set in terms of "Sandwich Matrix" and a congruence-free finite 0-simple semigroup S such that S is a semigroup of right translations of S and $V \cong S/J$ where J is an ideal of S consisting of some inner right translations of S; then we shall be able to obtain an ideal extension T of S by V and T will be a Δ-semigroup. We will mainly discuss the case where V is a 0-simple semigroup with zero divisors.

Let $V = M^0(I,G,M;P)$, where $P = (p_{ji})$, $p_{ji} \in G^0$, $j \in M$, $i \in I$ and $I = \{1,\ldots,n\}$, $M = \{1,\ldots,m\}$.

(5.1) *Construction of Permutation Group $\bar{G}$ with $\bar{G} \cong G$.*

The group $\bar{G}$ is not required to be transitive. Let k be a positive integer. Choose subgroups $H_1,\ldots,H_k$ of G such that the H_i's need not be distinct, but such that

$$(5.1.1) \qquad \bigcap_{\substack{1\leqslant i\leqslant k\\ x\in G}} x^{-1}H_i x = \{e\}.$$

Let G_{H_i} denote the permutation group of right translations by elements of G of the right cosets of H_i in G. (See Wieland [1964] or Rotman [1973b].) For simplicity, let G_{H_i} be a permutation group over a set X_i. The induced transitive representation $G \to G_{H_i}$ of G will be denoted by $h \mapsto h_i$, $h \in G$. Let $A = \bigcup_{i=1}^{k} X_i$ be the disjoint union and let A contain r elements. For each $h \in G$, define a permutation g_h over A by

$$xg_h = xh_i \text{ if } x \in X_i.$$

Then condition (5.1.1) yields that $h \to g_h$ is a faithful representation of G. Every faithful representation of G by permutations can be obtained in this manner. The representation thus obtained is denoted by $G \to \bar{G}$ and

$$\bar{G} = G_{H_1} \vee \ldots \vee G_{H_k}.$$

(5.2) *Transformation Semigroup Associated with V.*

Let $F : G \to \bar{G}$ be an isomorphism of G to the permutation group $\bar{G}$ of degree r. Choose sets, $A_1, \ldots, A_m$ (where $A_1 = A$), a set of mappings

$$\Sigma = \{\sigma_j : j \in M\},\ \sigma_j : A \to A_j$$

and a matrix $Q = (q_{ji})$, $j \in M$, $i \in I$, and possibly a set Π^Y of mappings such that the following conditions (5.2.1) through (5.2.6) hold.

(5.2.1) A_j satisfies:

(i) $|A_j| = r$ for all $j \in M$ where r is the degree of $\bar{G}$,

(ii) $A_{j_1} \neq A_{j_2}$ if $j_1 \neq j_2$; but A_{j_1} and A_{j_2} need not be disjoint.

(5.2.2) $\sigma_j : A \to A_j$ is a bijection for all $j \in M$.

In particular σ_1 is the identity map of A.

(5.2.3) If $p_{ji} \neq 0$ then $q_{ij} = F(p_{ji})$. If $p_{ji} = 0$, q_{ji} is a transformation of rank $\leqslant 2$ of A (that is, the range of q_{ji} is contained in $\{1, \ldots, r\}$ and has two or fewer elements).

(5.2.4) If $x, y \in A$ and $x\sigma_{j_1} = y\sigma_{j_2}$ then

$$xq_{j_1 i} = yq_{j_2 i} \text{ for all } i \in I.$$

Let Y be a finite set which may be empty but for which the following is satisfied.

(5.2.5) If there are $i_1, i_2, i_1 \neq i_2$, such that $q_{ji_1} = q_{ji_2}$ for all $j \in M$, then $Y \neq \emptyset$.

Let Π^Y be a set of mappings π_i^Y of Y into A. Then $\Pi^Y = \{\pi_i^Y : i \in I\}$ satisfies

(5.2.6) $\pi_{i_1}^Y \neq \pi_{i_2}^Y$ for i_1, i_2 satisfying (5.2.5).

Given Π^Y, $\bar{G}$, Σ, Q, let $D = Y \cup X_0$ where $X_0 = \bigcup_{j \in M} A_j$. For each $i \in I$, define $\pi_i : D \to A$ as follows:

$$x\pi_i = \begin{cases} x\sigma_j^{-1} q_{ji}, & x \in A_j \\ x\pi_i^Y \sigma_j^{-1} q_{ji}, x \in Y \text{ and } x\pi_i^Y \in A_j. \end{cases}$$

This is well defined because of (5.2.4).
Let $\Pi = \{\pi_i : i \in I\}$. Now define S to be the transformation semigroup generated by

$$\{\pi_i g \sigma_j : i \in I,\ g \in \bar{G},\ j \in M\}.$$

We call S a transformation semigroup associated with V and it is denoted by

$$S = T(\Sigma, \bar{G}, \Pi^Y, Q) \text{ or } T(V).$$

There are many possible choices for the construction of S, and $S = T(V)$ denotes any one of these. If $\Pi^Y = \emptyset$, S is denoted by

$$S = T(\Sigma, \bar{G}, Q).$$

Let I be the set of $f \in S$ of rank ≤ 2. Then I is an ideal of S and

$$\bar{S} = S/I \cong M^0(\Sigma, \bar{G}, \Pi; \bar{Q})$$

where

$$\bar{Q} = (\bar{q}_{ji}),\ \bar{q}_{ji} = \begin{cases} q_{ji}, & \text{if } q_{ji} \in G, \\ 0, & \text{otherwise.} \end{cases}$$

Thus $V = M^0(I, G, M; P) \cong M^0(\Sigma, \bar{G}, \Pi; Q)$
under $0 \to 0$ and $(i, x, j) \to \pi_i F(x) \sigma_j$.

We have written our operators above on the right so that $\pi_i\, g\, \sigma_j$ is obtained by first applying π_i, then g, then σ_j. We shall need the analogous construction obtained when our operators act on the left: $\sigma_i\, g\, \pi_j$ means first π_j, then g, then σ_i. If we need to specify "left" or "right" we use the notation:
for "left", $S^{(l)}$, $T^{(l)}(\Sigma^{(l)}, \bar{G}^{(l)}, \Pi^Y, Q^{(l)}, T^{(l)}(V), \bar{S}^{(l)}$;
for "right", $S^{(r)}$, $T^{(r)}(\Sigma^{(r)}, \bar{G}^{(r)}, \Pi^Y, Q^{(r)}), T^{(r)}(V), \bar{S}^{(r)}$.

To make left or right clearer, we let

$$\Sigma_I = \{\sigma_i : i \in I\},\ \Pi_M = \{\pi_j : j \in M\},$$

$$\Sigma_M = \{\sigma_j : j \in M\},\ \Pi_I = \{\pi_i : i \in I\},$$

$$Q^{(l)} = (q_{ji}),\ j \in M,\ i \in I,\ q_{ji} = \pi_j\sigma_i,$$

$$Q^{(r)} = (q_{ji}),\ j \in M,\ i \in I,\ q_{ji} = \sigma_j\pi_i.$$

Note

$$T^{(l)}(\Sigma^{(l)}, \bar{G}^{(l)}, \Pi^Y, Q^{(l)}) = \langle \{\sigma_i g \pi_j : i \in I,\ g \in G,\ j \in M\}\rangle,$$

$$V \cong M^0(\Sigma_I, \bar{G}, \Pi_M; Q^{(l)}),\ \text{under } (i,x,j) \to \sigma_i F(x)\pi_j,\ 0 \to 0,$$

$$T^{(r)}(\Sigma^{(r)}, \bar{G}^{(r)}, \Pi^Y, Q^{(r)}) = \langle \{\pi_i g \sigma_j : i \in I,\ g \in G,\ j \in M\}\rangle,$$

$$V \cong M^0(\Pi_I, \bar{G}, \Sigma_M; Q^{(r)}),\ \text{under } (i,x,j) \to \pi_i F(x)\sigma_j,\ 0 \to 0,$$

where $\langle X \rangle$ denotes the semigroup generated by a set X.

We have assumed V is a finite 0-simple semigroup with zero divisors. But, with a slight change, the above holds for the case where V is a finite 0-simple semigroup without zero divisors.

PROPOSITION 5.3. *Given a finite 0-simple semigroup V, there is a transformation semigroup S over a finite set such that $S/I \cong V$ where elements of I have rank $\leqslant 2$. (If V has no zero divisor, disregard I.) Every S can be obtained in the above-mentioned manner.*

Let S be a congruence-free semigroup, $S = M^0(I,\{1\},M;P)$, $|I| = n$, $|M| = m$. Let c_i be the i-th column vector of length m of P,

$$c_i = \begin{pmatrix} p_{1i} \\ \vdots \\ p_{mi} \end{pmatrix} \quad i = 1,\ldots,n.$$

Then

$$P = (c_1,\ldots,c_n).$$

If f is a transformation $M \to M^0$, we define $f\cdot c_i$ by

$$f\cdot c_i = \begin{pmatrix} p_{1f,i} \\ p_{2f,i} \\ \vdots \\ p_{mf,i} \end{pmatrix}$$

Let P_c be the set of column vectors of P, $P_c = \{c_1,\ldots,c_n\}$, and let 0 be the zero vector. Since S is congruence-free, every bitranslation of S is determined by a pair (q,f) satisfying $p_{jf,i} = p_{j,qi}$ for all i,j. (See lemma 4.1.). The set of (q,f) is denoted by $\Omega^*(S)$. Then we have

$$fP_c \subseteq P_c \cup \{0\} \text{ if and only if } (q,f) \in \Omega^*(S) \text{ for some } q.$$

Let T_m be the set of all vectors of length m whose entries are either 1 or 0, and let C be a subset of $T_m \setminus \{0\}$. Let X be a set of transformations over the set M^0, $|M| = m$. Let $XC = C \cup \{fc : f \in X, c \in C\}$. If $XC \subseteq C \cup \{0\}$ we say C is X-closed. Let $\bar{X}$ denote the semigroup generated by X. Then $\bar{X}C = \bigcup_{i=0}^{\infty} X^i C$ where $X^0 c = c$ for all $c \in C$ and $X^i C = X^{i-1}(XC)$. Because of finiteness, $\bigcup^{\infty}$ becomes a finite union. Also C is X-closed if and only if C is $\bar{X}$-closed; $C \to XC$ is a closure operation in T_m.

Let $C = \{c_1,\ldots,c_l\}$ where $c_1,\ldots,c_l$ are non-zero and distinct. If C is $\bar{X}$-closed and C composes a $m \times l$ sandwich matrix $(c_1,\ldots,c_l)$, that is, no row vector is 0 and all rows are distinct (this property is independent of the order $c_1,\ldots,c_l$, but it is a property of C), then C is called *satisfactory*.

THEOREM 5.4. *Let V be a finite 0-simple Δ-semigroup. There is a finite congruence-free 0-simple semigroup S and an ideal extension T of S by V such that T is a Δ-semigroup.*

Proof. By proposition 5.3 we can find a transformation semigroup, say F (written instead of S over a finite set M^0 such that $|M| = m$, and $V^* = V\setminus\{0\}$ can be mapped by a partial isomorphism into F. It is easy to see that T_m is satisfactory. However we explain how to find all satisfactory C's. Suppose $\bar{F}C$ is not satisfactory. If the matrix composed by $\bar{F}C$ contains row zero vectors $r_{j_1},\ldots,r_{j_s}$, then add to C a column vector c whose j_1-th,...,j_s-th entries are 1, so that $\bar{F}(C \cup \{c\})$ has no row zero vectors. Assume that $\bar{F}C = \{c_1,\ldots,c_l\}$ and $c_i = \begin{pmatrix} p_{1i} \\ \vdots \\ p_{mi} \end{pmatrix}$, $i = 1,\ldots,l$,

and that $\bar{F}C$ has no row zero vectors. We define an equivalence λ on M as follows:

$$j_1 \lambda j_2 \text{ if and only if } p_{j_1 i} = p_{j_2 i} \text{ for } i = 1,\ldots,l.$$

Let m_λ denote the maximum of the numbers of elements of λ-classes. Then, add to $\bar{F}C$ k new column vectors, $d_1,\ldots,d_k$ such that $FC \cup F\{d_1,\ldots,d_k\}$ is satisfactory where k satisfies $2^{k-1} \leqslant m_\lambda - 1 < 2^k$. If FC is satisfactory then $\bar{F}C \cup \bar{F}C'$ is satisfactory for all C' where C' consists of distinct column vectors and every one of C' is different from every one of C. Therefore we can say as follows: For any non-zero $c \in T_m$ there is a sequence $c_1,\ldots,c_t \in T_m$ $(t > 1)$ such that $\bar{F}c_1$, $\bar{F}c_1 \cup \bar{F}c_2,\ldots, \bigcup_{i=1}^{t-1} \bar{F}c_i$ are not satisfactory but $\bigcup_{i=1}^{t} \bar{F}c_i$ is satisfactory. After all, let P be an $m \times l$ matrix $P = (c_1,\ldots,c_l)$ such that $\{c_1,\ldots,c_l\}$ is satisfactory. (If $C = T_m$ then P is $m \times (2^m-1)$.) Assume $S = M^0(I_S,\{1\},M;P)$ and the set $\{c_1,\ldots,c_l\}$of P is F-closed. For each $f \in F$ there is a unique i', $1 \leqslant i' \leqslant l$, such that $fc_i = c_{i'}$. Define $q : I^0 \to I^0$ by $qi = i'$, $q0 = 0$. Thus q is uniquely determined such that $(q,f) \in \Omega^*(S)$. Let Q denote the set of q. Permutability of right and left translations of S are automatically satisfied because $S^2 = S$ [Petrich 1973a]. Therefore an extension T of S by V can be obtained. By corollary 2.4, T is a Δ-semigroup.

Example 1. Let $V = M^0(I,\{1\},M;P)$ where $|I| = |M| = 2$ and $P = \begin{pmatrix} 1 & 1 \\ 0 & 0 \end{pmatrix}$. Then $V = \{0,(1,1),(1,2),(2,1),(2,2)\}$. Let $\bar{G} = \left\{\begin{pmatrix} 1 & 2 \\ 1 & 2 \end{pmatrix}\right\}$. Then $F = T^{(r)}(V)$ is generated by $\{f_{11},f_{12},f_{21},f_{22}\}$:

$\Pi^{(r)}$ \ $\Sigma^{(r)}$	$\pi_1^{(r)}$ $\begin{pmatrix} 1 & 2 & 3 & 4 \\ 1 & 2 & 1 & 1 \end{pmatrix}$	$\pi_2^{(r)}$ $\begin{pmatrix} 1 & 2 & 3 & 4 \\ 1 & 2 & 2 & 2 \end{pmatrix}$
$\sigma_1^{(r)} = \begin{pmatrix} 12 \\ 12 \end{pmatrix}$	$\begin{pmatrix} 12 \\ 12 \end{pmatrix}$	$\begin{pmatrix} 12 \\ 12 \end{pmatrix}$
$\sigma_2^{(r)} = \begin{pmatrix} 12 \\ 13 \end{pmatrix}$	$\begin{pmatrix} 12 \\ 11 \end{pmatrix}$	$\begin{pmatrix} 12 \\ 12 \end{pmatrix}$

$$f_{11} = \begin{pmatrix} 1 & 2 & 3 & 4 \\ 1 & 2 & 1 & 1 \end{pmatrix}$$

$$f_{12} = \begin{pmatrix} 1 & 2 & 3 & 4 \\ 1 & 3 & 1 & 1 \end{pmatrix}$$

$$f_{21} = \begin{pmatrix} 1 & 2 & 3 & 4 \\ 1 & 2 & 2 & 2 \end{pmatrix}$$

$$f_{22} = \begin{pmatrix} 1 & 2 & 3 & 4 \\ 1 & 3 & 3 & 3 \end{pmatrix}$$

Let $S = M^0(I_S, \{1\}, M; P_S)$ where

$$P_S = \begin{pmatrix} 1 & 0 & 1 & 1 & 0 & 1 \\ 0 & 1 & 0 & 0 & 1 & 1 \\ 0 & 0 & 1 & 1 & 1 & 1 \\ 0 & 0 & 0 & 1 & 1 & 1 \end{pmatrix}$$ such that the set of column vectors is F-closed.

Then

$K = T^{(l)}(V)$ is generated by $\{k_{11}, k_{12}, k_{21}, k_{22}\}$:

$Q^{(l)}$ — $\Sigma^{(l)}$ \ $\Pi^{(l)}$	$\sigma_1^{(l)}$ $\begin{pmatrix} 4 & 2 & 6 \\ 4 & 2 & 6 \end{pmatrix}$	$\sigma_2^{(l)}$ $\begin{pmatrix} 4 & 2 & 6 \\ 1 & 5 & 6 \end{pmatrix}$
$\pi_1^{(l)} = \begin{pmatrix} 1 & 2 & 3 & 4 & 5 & 6 \\ 4 & 2 & 4 & 4 & 2 & 6 \end{pmatrix}$	$\begin{pmatrix} 4 & 2 & 6 \\ 4 & 2 & 6 \end{pmatrix}$	$\begin{pmatrix} 4 & 2 & 6 \\ 4 & 2 & 6 \end{pmatrix}$
$\pi_2^{(l)} = \begin{pmatrix} 1 & 2 & 3 & 4 & 5 & 6 \\ 4 & 0 & 6 & 6 & 2 & 6 \end{pmatrix}$	$\begin{pmatrix} 4 & 2 & 6 \\ 6 & 0 & 6 \end{pmatrix}$	$\begin{pmatrix} 4 & 2 & 6 \\ 4 & 2 & 6 \end{pmatrix}$

$$k_{11} = \begin{pmatrix} 1 & 2 & 3 & 4 & 5 & 6 \\ 4 & 2 & 4 & 4 & 2 & 6 \end{pmatrix} \qquad k_{12} = \begin{pmatrix} 1 & 2 & 3 & 4 & 5 & 6 \\ 4 & 0 & 6 & 6 & 2 & 6 \end{pmatrix}$$

$$k_{21} = \begin{pmatrix} 1 & 2 & 3 & 4 & 5 & 6 \\ 1 & 5 & 1 & 1 & 5 & 6 \end{pmatrix} \qquad k_{22} = \begin{pmatrix} 1 & 2 & 3 & 4 & 5 & 6 \\ 1 & 0 & 6 & 6 & 5 & 6 \end{pmatrix}.$$

An ideal extension T of S by V is determined by V, F, S, K.

Finally we comment on the ideal extension T of a given congruence-free S by a 0-simple semigroup V. All semigroups are finite.

Let $\Omega(S)$ denote the translational hull of S. By propositions 2.2, 2.3 and corollary 2.4, we have

PROPOSITION 5.5. *Let S be a congruence-free 0-simple semigroup. If T is a subsemigroup of $\Omega(S)$ and S is an ideal of T such that T/S is a 0-simple Δ-semigroup, then T is a Δ-semigroup. Every Δ-semigroup T which is an ideal extension of S by a 0-simple Δ-semigroup V can be obtained in this manner.*

Let $\Omega_0(S)$ denote the component containing S in the greatest semilattice decomposition of $\Omega(S)$. The following refines proposition 5.5. In the following, "0-simple Δ-semigroup V" admits "a null semigroup of order 2."

THEOREM 5.6. *Let S be a congruence-free 0-simple semigroup. There exists a 0-simple Δ-semigroup V and a Δ-semigroup T which is an ideal extension of S by V if and only if one of the following holds. Let $V^* = V\setminus\{0\}$.*

(5.6.1) *S is arbitrary. Then choose a Δ-group V^* which is isomorphic to a subgroup of $\Omega(S)/S$.*

(5.6.2) *$\Omega(S)/S$ has a 0-simple principal factor which is not an inverse semigroup. Then choose a right or left zero subsemigroup V^* of order 2 of $\Omega(S)/S$.*

(5.6.3) *$\Omega_0(S) \neq S$. Then choose a null subsemigroup V of order 2 of $\Omega_0(S)/S$.*

(5.6.4) *$\Omega_0(S) \neq S$ and $\Omega_0(S)/S$ has a 0-simple ideal I. Then let $V \cong I$.*

Example 2. Let $S = M^0(I,\{1\},M;P)$ where P is defined by

$$P = \begin{pmatrix} 1 & 1 & 0 \\ 1 & 0 & 1 \\ 0 & 1 & 1 \end{pmatrix}.$$

Then there is no Δ-semigroup which is an ideal extension of S by any 0-simple semigroup with zero divisors. But there exists a Δ-semigroup which is an ideal extension of S by a group with zero adjoined.

Acknowledgement

The author wants to express heartfelt thanks to the organising committee of this Monash Conference for its invitation to present a paper and also thanks to Dr P.G. Trotter for presenting the paper on behalf of the author.

References

1956 T. Tamura, Indecomposable completely simple semigroups except groups. *Osaka Math. J.*, 8 (1956), 35-42.

1960 T. Tamura, Decompositions of a completely simple semigroup. *Osaka Math. J.*, 12 (1960), 269-275.

1961 A.H. Clifford and G.B. Preston, *The algebraic theory of semigroups, Vol. 1*, Math. Surveys of the Amer. Math. Soc. 7, 1961.

1964 H. Wielandt, *Finite permutation groups*. Academic Press, 1964.

1966 R.J. Warne, Extensions of completely 0-simple semigroups by completely 0-simple semigroups. *Proc. Amer. Math. Soc.*, 17 (1966), 524-526.

1967a A.H. Clifford and G.B. Preston, *The algebraic theory of semigroups, Vol. 2,* Math. Surveys of the Amer. Math. Soc., 7, 1967.

1967b M. Petrich, Congruences on extensions of semigroups. *Duke Math. J.,* 34 (1967), 215-224.

1969a B.M. Schein, Commutative semigroups where congruences form a chain. *Bull. de l'Acad. Polon. des Sciences,*9 (1969),523-527.

1969b T. Tamura, Commutative semigroups whose lattice of congruences is a chain. *Bull. Soc. Math. France,* 97 (1969), 369-380.

1970 W.A. Etterbeek, *Semigroups whose lattice of congruences forms a chain.* Dissertation, University of California, Davis, 1970.

1973a M. Petrich, *Introduction to semigroups.* Merrill, Columbus, Ohio, 1973.

1973b J.J. Rotman, *The theory of groups.* Allyn and Bacon, Boston, 1973.

1974 G. Lallement, A note on congruences on Rees matrix semigroups. *Semigroup Forum,* 8 (1974), 89-92.

1976 P.G. Trotter, Exponential Δ-semigroups. *Semigroup Forum,* 12 (1976), 313-331.

1977 P.G. Trotter and T. Tamura, Completely semisimple inverse Δ-semigroups admitting principal series. *Pac. J. Math.,* 68 (1977), 515-525.

1978 K.G. Johnston, Congruence lattices of Rees matrix semigroups. *Semigroup Forum,* 15 (1978), 247-261.

1979a C. Bonzini and A. Cherubini Spoletini, Sugli *E-m* semigruppi. *Università Degli Studi de Milano* (Preprint), 1979.

1979b T. Tamura, Finite semigroups whose congruences form chains. *Proceedings of the Conference on Semigroups in honour of A.H. Clifford,* 1978, Tulane University (1979), 94-116.

DIRECT PRODUCTS OF CYCLIC SEMIGROUPS

D. C. Trueman

Department of Mathematics, Monash University, Clayton, Victoria, Australia, 3168.

Abstract

It will be proved that a finite semigroup which is decomposable into a direct product of cyclic semigroups which are not groups is uniquely so decomposable. It will then be determined when a finite semigroup has such a decomposition and how its non-group cyclic direct factors, if they exist, can be found.

Preliminaries

Let $S_1,\ldots,S_n$ be finite cyclic semigroups, each S_i having index r_i and period m_i, $1 \leqslant i \leqslant n$. Suppose that x_i generates $S_i = \langle x_i \rangle$, K_i being the maximal subgroup of S_i with identity e_i. Denote the (external) direct product of $S_1,\ldots,S_n$ by $S = S_1 \times \ldots \times S_n$

$$= \{(x_1^a,\ldots,x_n^b) \mid x_i^s \in S_i,\ s = g_i,\ 1 \leqslant g_i \leqslant r_i + m_i - 1,\ 1 \leqslant i \leqslant n\}$$

with product defined by

$$(x_1^a,\ldots,x_n^b)(x_1^c,\ldots,x_n^d) = (x_1^a x_1^c,\ldots,x_n^b x_n^d).$$

Without loss of generality, assume that $r_h \leqslant r_i$ for each h,i such that $1 \leqslant h < i \leqslant n$, and if $r_i = \ldots = r_j$ for some i and j, $1 \leqslant i < j \leqslant n$, suppose $m_i \geqslant \ldots \geqslant m_j$. Denote the order of S_i by w_i, and the power of the ith component in $(x_1^a,\ldots,x_n^b)$, by g_i. For definitions of terms used in this paper, see Clifford and Preston [1961 and 1967]. For further results and fuller proofs see (Trueman [A]).

ISBN 0-12-319450-4

1. Properties

S is archimedean, contains a unique idempotent $e = (e_1,\ldots,e_n)$ and a unique maximal subgroup $H_e = K_1 \times \ldots \times K_n$.

THEOREM 1.1. *A finite cyclic semigroup which is not a group cannot be decomposed into a direct product of cyclic semigroups, with at least two of the direct factors non-trivial.*

Proof. Let C be a finite cyclic semigroup of index greater than 1, and suppose that $C \cong S_1 \times \ldots \times S_p$, a direct product of non-trivial cyclic semigroups. Let $(x_1^a,\ldots,x_p^b) \in S_1 \times \ldots \times S_p$ generate $S_1 \times \ldots \times S_p$. Then there exists a positive integer t such that $(x_1^a,\ldots,x_p^b)^t = (x_1,\ldots,x_p)$. Since at least one of the direct factors is not a group, then there exists a k, $1 \leqslant k \leqslant p$, such that $g_k t = 1$, so $t = 1$. Hence, $(x_1,\ldots,x_p)$ generates $S_1 \times \ldots \times S_p$. So, for each i, $1 \leqslant i \leqslant p$, there exists a positive integer $s = u_i$, greater than 1, such that $(x_1,\ldots,x_{i-1},x_i,x_{i+1},\ldots,x_p)^s = (x_1,\ldots,x_{i-1},x_i^2,x_{i+1},\ldots,x_p)$. Hence, for each j, $j \neq i$, S_j is a group, and either S_i is a group or its index is 2. But this is true for each i, so all the direct factors are groups, which conflicts with our hypotheses.

THEOREM 1.2. *Let each S_i have index r_i, greater than 1, $1 \leqslant i \leqslant n$. Then $S = S_1 \times \ldots \times S_n$ has a unique minimal set of generators containing $\prod_{i=1}^{n} (r_i + m_i - 1) - \prod_{i=1}^{n} (r_i + m_i - 2)$ elements.*

Proof. Consider the set X of elements that, for some i, have component x_i, i.e. $X = S \setminus S^2$. Then any element of S is a product of such elements, and X is contained in all sets generating S, and is therefore its unique minimal set of generators with

$$\prod_{i=1}^{n} (r_i + m_i - 1) - \prod_{i=1}^{n} (r_i + m_i - 2) \text{ elements.}$$

A unique minimal set of generators of a semigroup C is henceforth called the *basis* of C.

For any two positive integers s_1, s_2, denote their greatest common divisor by $s_1 \wedge s_2$ and their least common multiple by $s_1 \vee s_2$.

THEOREM 1.3. *Let $(x_1^a,\ldots,x_n^b) \in S$, and let t be the least integer q such that, for each i simultaneously, $1 \leqslant i \leqslant n$, $g_i q \leqslant r_i$. Then*

$(x_1^a,\dots,x_n^b)$ has index t and period $\varepsilon = p_1 \vee \dots \vee p_n$, where $p_i = m_i/(g_i \wedge m_i)$, $1 \leqslant i \leqslant n$.

Proof. For each i, $1 \leqslant i \leqslant n$, the ith component x_i^s generates a cyclic subsemigroup of S_i of index t_i, where t_i is the least integer q_i such that $sq_i \geqslant r_i$, and period $p_i = m_i/(g_i \wedge m_i)$. Hence, the index of $(x_1^a,\dots,x_n^b)$ is q, as above, and its period is the least common multiple of $p_1,\dots,p_n$.

THEOREM 1.4. *If at least one of the direct factors of S is not a group and at least two are non-trivial, then S contains a proper maximal cyclic subsemigroup which is not properly contained in any cyclic subsemigroup of S and is of maximal index r_n and maximal period $m_1 \vee \dots \vee m_n$ with respect to the other cyclic subsemigroups of S.*

Proof. Any element $(x_1^a,\dots,x_n^b)$ of S has index t, where $t \leqslant r_n$, and period ε, where $\varepsilon = p_1 \vee \dots \vee p_n \leqslant m_1 \vee \dots \vee m_n$, and $p_i = m_i/(g_i \wedge m_i)$, $1 \leqslant i \leqslant n$. Moreover, $\langle (x_1,\dots,x_n)\rangle$ has index r_n and period $m_1 \vee \dots \vee m_n$, and, by theorem 1.1, is properly contained in S.

2. The decomposition of S

Let $T = T_1 \times \dots \times T_l$ be a direct product of non-trivial finite cyclic semigroups, where $T_j = \langle y_j \rangle$ has index r_j' and period m_j', and the maximal subgroup of T_j has identity f_j, $1 \leqslant j \leqslant l$. Suppose $r_j' \leqslant r_k'$ for each j,k such that $1 \leqslant j < k \leqslant l$, and if $r_i' = \dots = r_j'$ for some i and j, $1 \leqslant i < j \leqslant l$, suppose $m_i' \geqslant \dots \geqslant m_j'$. Denote the order of T_j by w_j', $1 \leqslant j \leqslant l$.

THEOREM 2.1. *Let $S \cong T$, where $T = T_1 \times \dots \times T_l$ is a direct product of non-trivial cyclic semigroups, each S_i having index $r_i > 1$, $1 \leqslant i \leqslant n$. Then $l = n$ and $S_i \cong T_i$ for each i, $1 \leqslant i \leqslant n$.*

Proof. If $n = 1$, then $l = 1$, by theorem 1.1. Let $n > 1$. Then S contains a proper cyclic subsemigroup of maximal index r_n and maximal period, by theorem 1.4. Also, T contains a proper maximal cyclic subsemigroup of maximal index r_l' and maximal period. Since $S \cong T$, then $r_n = r_l'$.

Any element of S of index r_n with minimal period, has period m_n, and any element of T of index $r_l' = r_n$ with minimal period, has period

m'_l. But $S \cong T$, so $m_n = m'_l$ and $S_n \cong T_l$.

If $r_{n-1} = r_n$, then $r'_{l-1} \nless r_{n-1} = r'_l$. If $r_{n-1} < r_n$, suppose that $r_{n-1} < r'_{l-1}$. S contains exactly $w_1 \ldots w_{n-1} d_n$ elements of index r'_{l-1}, where d_n is the number of elements in S_n of index r'_{l-1}. There are $w'_1 \ldots w'_{l-1} d'$ elements in T of index r'_{l-1}, where d' is the number of elements in T_l of index r'_{l-1}, plus at least one more, for example, $(f_1, \ldots, f_{l-2}, y_{l-1}, f_l)$. Since $S \cong T$ and $S_n \cong T_l$, then $d_n = d'$, $w_n = w'_l$ and $w'_1 \ldots w'_{l-1} d' = w_1 \ldots w_{n-1} d_n$, a contradiction, since S and T must contain the same number of elements of index r'_{l-1}. Hence, $r_{n-1} \nless r'_{l-1}$. Using a similar argument, $r'_{l-1} \nless r_{n-1}$. So, $r_{n-1} = r'_{l-1}$.

If $r_{n-1} = r_n$ and $m_{n-1} = m_n$, then $m'_{l-1} \nless m_{n-1}$. So, suppose that $r_{n-1} < r_n$ or, if $r_{n-1} = r_n$, that $m_{n-1} > m_n$. Suppose that $m_{n-1} > m'_{l-1}$. S contains exactly $q_1 \ldots q_{n-1} z_n$ elements of index r_{n-1} and period dividing m'_{l-1}, where q_u is the number of elements in S_u of period dividing m'_{l-1}, $1 \leqslant u \leqslant n$, and z_n is the number of elements in S_n of index r_{n-1} and period dividing m'_{l-1}. T contains $q'_1 \ldots q'_{l-1} z'$ such elements, where q'_h is the number of elements in T_h of period dividing m'_{l-1} and z' is the number of elements in T_l of index r'_{l-1} and period dividing m'_{l-1}, plus at least one more, for example $(f_1, \ldots, f_{l-2}, y_{l-1}, f_l)$ Since $S \cong T$ and $S_n \cong T_l$, then $z_n = z'$, $q_n = q'_l$ and $q_1 \ldots q_n = q'_1 \ldots q'_l$, since the number of elements in S of period dividing m'_{l-1} must equal the number of such elements in T. Hence, $q_1 \ldots q_{n-1} z_n = q'_1 \ldots q'_{l-1} z'$, a contradiction, since S and T must contain the same number of elements of index r_{n-1} and period dividing m'_{l-1}. Thus $m_{n-1} \ngtr m'_{l-1}$. Using a similar argument, $m'_{l-1} \ngtr m_{n-1}$. So $m'_{l-1} = m_{n-1}$ and $S_{n-1} \cong T_{l-1}$.

The theorem follows by induction, assuming $S_{n-k} \cong T_{l-k}, \ldots, S_n \cong T_l$ for some k, $1 \leqslant k \leqslant n - 1$, and considering $S_{n-(k+1)}$ and $T_{l-(k+1)}$, and the number of elements in S of indices $r_{n-(k+1)}$ or $r'_{l-(k+1)}$ and periods dividing $m_{n-(k+1)}$ or $m'_{l-(k+1)}$.

3. Determining non-group cyclic direct factors

Let C be a finite semigroup. If C can be decomposed into a direct product of cyclic semigroups which are not groups, then $C \cong S = S_1 \times \ldots \times S_n$, say, where each S_i is cyclic of index $r_i > 1$ and period m_i, $1 \leqslant i \leqslant n$. Hence, C must be a commutative, archimedean, unipotent, non-cyclic semigroup, and $C \setminus C^2$ must be a basis of C. Suppose that C is such a semigroup. To find its possible direct factors:

(a) Find the maximal index of the elements of C (or $C\setminus C^2$), and call this r_n. Then find the minimal period of the elements of C (or $C\setminus C^2$) with index r_n, and call this m_n. If $C \cong S$, then S_n has index r_n and period m_n.

(b) If $r_{n-1} < r_n - j$ for some j, $j = 0,1,2,\ldots,r_n-2$, then C must contain $w_1\ldots w_{n-1}d_n$ elements of index $r_n - j$, where d_n is the number of elements in S_n of index $r_n - j$. Check this for $j = 0,1,2,\ldots,$ until a contradiction results for some $r_n - i$. Then let $r_n - i = r_{n-1}$, for otherwise, $C \not\cong S$.

(c) Find the minimal period, say t_{n-1}, of the elements of C with index r_{n-1}. If $m_{n-1} \geqslant t_{n-1} + j$ for some $j \geqslant 0$, then C must contain exactly $q_1\ldots q_{n-1}z_n$ elements of index r_{n-1} and period dividing $t_{n-1}+j$, where q_u is the number of elements in S_u of period dividing $t_{n-1} + j$, $1 \leqslant u \leqslant n$, and z_n is the number of elements in S_n of index r_{n-1} and period dividing $t_{n-1} + j$; $q_1\ldots q_{n-1}$ is the number of elements in S of period dividing $t_{n-1} + j$ divided by q_n, the number of elements in S_n of period dividing $t_{n-1} + j$. Check this for $j = 0,1,2,\ldots,$ until a contradiction results for some $t_{n-1} + k$. Then let $t_{n-1} + k = m_{n-1}$, for otherwise, $C \not\cong S$.

Using similar arguments, $r_{n-2} = r_{n-1} - i$ for some i, $0 \leqslant i \leqslant r_{n-1} - 2$, and $m_{n-2} = t_{n-2} + k$ for some k, $k \geqslant 0$. Continue in this manner until all possible indices and periods of the possible direct factors have been found, or until a contradiction results, and it follows that $C \not\cong S$.

(d) Check that C has order $\prod_{i=1}^{n} (r_i + m_i - 1)$, $C\setminus C^2$ has order $\prod_{i=1}^{n} (r_i + m_i - 1) - \prod_{i=1}^{n} (r_i + m_i - 2)$, and the maximal period of the elements of C is $m_1 \vee \ldots \vee m_n$, for otherwise, $C \not\cong S$.

(e) Label the distinct elements of $C\setminus C^2$, say $c_1,\ldots,c_p$, corresponding to the elements of $S\setminus S^2$, in a one-to-one manner, as follows.

$$c_k = x(a,\ldots,b) \text{ if } \langle c_k\rangle \cong \langle (x_1^a,\ldots,x_n^b)\rangle, \text{ where } (x_1^a,\ldots,x_n^b) \in S\setminus S^2, \quad 1 \leqslant k \leqslant p,$$

with product defined as

$$x(a,\ldots,b)x(c,\ldots,d) = x(s,\ldots,t),$$

where $s = a + c$, if $a + c < r_1$, and $s = (a + c - r_1) \bmod m_1 + r_1$, otherwise, with similar rules for the other components.

If, for one such labeling, product in C corresponds to that in S, then $C \cong S$. If not, then $C \not\cong S$.

4. Direct products of finite monogenic inverse semigroups

Let $U = U_1 \times \ldots \times U_n$ be a direct product of finite monogenic inverse semigroups, (a monogenic inverse semigroup being a semigroup I such that, for some $a \in I$, I is generated by a and a^{-1}), where $U_i = \langle u_i, u_i^{-1} \rangle$, $u_i u_i^{-1} u_i = u_i$, $u_i^{-1} u_i u_i^{-1} = u_i^{-1}$, and u_i has index r_i and period m_i, $1 \leqslant i \leqslant n$. Suppose that $r_h \leqslant r_i$ for each h,i such that $1 \leqslant h < i \leqslant n$, and if $r_i = \ldots = r_j$ for some i and j, $1 \leqslant i < j \leqslant n$, suppose $m_i \geqslant \ldots \geqslant m_j$. When we write $U \cong V$, where $V = V_1 \times \ldots \times V_l$ is a direct product of monogenic inverse semigroups, let $V_j = \langle v_j, v_j^{-1} \rangle$, where v_j has index r'_j and period m'_j. Suppose $r'_j \leqslant r'_k$ for each j,k such that $1 \leqslant j < k \leqslant l$, and if $r'_i = \ldots = r'_j$ for some i and j, $1 \leqslant i < j \leqslant l$, suppose $m'_i \geqslant \ldots \geqslant m'_j$.

THEOREM 4.1. *Let $W = \langle u, u^{-1} \rangle$ be a monogenic inverse semigroup, where u has index r, greater than 1, and period m. Then any element of W has index less than or equal to r and period dividing m. If an element of W has index r, it has period m.*

Proof. Let $a \in W$. Then, for some k, $1 \leqslant k \leqslant r + m - 1$,

$$a = u^{\pm k} ,$$

or, for some i and j, $1 \leqslant i < r$, $1 \leqslant j < r$,

$$a = \begin{cases} u^i u^{-j} , \text{ or} \\ u^{-i} u^j , \end{cases}$$

or, for some i, j and k, such that $1 \leqslant i < r$, $1 \leqslant j < r$, and $1 \leqslant k < r$, with $j > i$ and $j > k$,

$$a = \begin{cases} u^i j^{-j} u^k , \text{ or} \\ u^{-i} u^j u^{-k}. \end{cases}$$

In each case, it can be shown that $\langle a \rangle$ has the required index and period.

THEOREM 4.2. *Let* $U = U_1 \times \ldots \times U_n$ *be a direct product of finite monogenic inverse semigroups, with* r_i *greater than 1 for at least one* i, $1 \leqslant i \leqslant n$, *and at least two of the direct factors non-trivial. Then* U *is a non-commutative, non-monogenic inverse semigroup containing subsemigroups of maximal index* r_n *and maximal period* $\varepsilon = m_1 \vee \ldots \vee m_n$.

Proof. Let $r_i > 1$. If U is commutative, then $(u_1,\ldots,u_i,\ldots,u_n)(u_1^{-1},\ldots,u_i^{-1},\ldots,u_n^{-1}) = (u_1^{-1},\ldots,u_i^{-1},\ldots,u_n^{-1})(u_1,\ldots,u_i,\ldots,u_n)$, so $u_iu_i^{-1} = u_i^{-1}u_i$ and $u_i = u_i^{-1}u_i^2$. Then u_i^{s-1}, where $s = r_i$, is an element of the maximal subgroup of $\langle u_i \rangle$, a contradiction. Hence, U is non-commutative.

Suppose that $U = \langle a, a^{-1} \rangle$, where $a \in U$ has index t and period ε. Since the maximal subgroup of $\langle a \rangle$ equals that of $\langle a^{-1} \rangle$, then the order of $\langle a, a^{-1} \rangle$ is less than or equal to

$$2(r_n - 1) + m_1 \vee \ldots \vee m_n + 2(r_{n-1} - 1)^2 + b,$$

where b is the number of elements $a^ia^{-j}a^k$, with $1 \leqslant i < t$, $1 \leqslant j < t$, $1 \leqslant k < t$, and $j > i$, $j > k$. But U_n has order

$$2(r_n - 1) + m_n + 2(r_n - 1)^2 + c,$$

where c is the number of elements $u_n^iu_n^{-j}u_n^k$, with $1 \leqslant i < r_n$, $1 \leqslant j < r_n$, $1 \leqslant k < r_n$ and $j > i$, $j > k$. Then, if the order of $U_1 \times \ldots \times U_{n-1}$ is p, then U has order $(2(r_n - 1) + m_n + 2(r_n - 1)^2 + c)p > 2(r_n - 1) + 2(r_n - 1)^2 + b + m_1 \ldots m_n \geqslant 2(r_n - 1) + 2(r_n - 1)^2 + b + m_1 \vee \ldots \vee m_n$, a contradiction. Hence, U is non-monogenic.

Clearly, $\langle (u_1,\ldots,u_n) \rangle$ has maximal index r_n and maximal period $m_1 \vee \ldots \vee m_n$.

THEOREM 4.3. *Let* $U \cong V$, *where* $V = V_1 \times \ldots \times V_l$ *is a direct product of non-trivial monogenic inverse semigroups, each* U_i *having index* $r_i > 1$, $1 \leqslant i \leqslant n$. *Then* $l = n$ *and* $U_i \cong V_i$, $1 \leqslant i \leqslant n$.

Proof. If $n = 1$, then $l = 1$, by theorem 4.2. U contains cyclic subsemigroups with maximal index r_n and maximal period, and V contains cyclic subsemigroups with maximal index r_l' and maximal period. Since

$U \cong V$, then $r_n = r'_l$. Any cyclic subsemigroup of U of index r_n with minimal period, has period m_n, and any cyclic subsemigroup of V of index $r'_l = r_n$ and minimal period, has period m'_l. So $m_n = m'_l$. Since $\langle u_n, u_n^{-1}\rangle \cong \langle v_l, v_l^{-1}\rangle$ if and only if $\langle u_n\rangle \cong \langle v_l\rangle$, then $U_n \cong V_l$.

Suppose $r_{n-1} < r'_{l-1}$. Then U contains exactly $w_1 \ldots w_{n-1} d_n$ elements of index r'_{l-1}, where w_i is the order of U_i, $1 \leqslant i \leqslant n$, and d_n is the number of elements in U_n of index r'_{l-1}. V contains $w'_1 \ldots w'_{l-1} d'$ such elements, where w'_j is the order of V_j, $1 \leqslant j \leqslant l - 1$, and d' is the number of elements in V_l of index r'_{l-1}. V also contains more of these, for example $(v_1 v_1^{-1}, \ldots, v_{l-2} v_{l-2}^{-1}, v_{l-1} v_l v_l^{-1})$, a contradiction. So $r_{n-1} \not< r'_{l-1}$, and using a similar argument, $r'_{l-1} \not< r_{n-1}$. Thus, $r_{n-1} = r'_{l-1}$.

Suppose $m_{n-1} > m'_{l-1}$. The number of elements in U with index r_{n-1} and period dividing m'_{l-1}, is $q_1 \ldots q_{n-1} z_n$, where q_i is the number of elements in U_i with period dividing m'_{l-1}, and z_n is the number of elements in U_n of index r_{n-1} and period dividing m'_{l-1}. V contains $q'_1 \ldots q'_{l-1} z'$ such elements, where q'_j is the number of elements in V_j of period dividing m'_{l-1}, and z' is the number of elements in V_l of index r'_{l-1} and period dividing m'_{l-1}, plus more, for example, $(v_1 v_1^{-1}, \ldots, v_{l-2} v_{l-2}^{-1}, v_{l-1}, v_l v_l^{-1})$, a contradiction. So, $m_{n-1} \not> m'_{l-1}$, and using a similar argument, $m'_{l-1} \not> m_{n-1}$. Hence, $m_{n-1} = m'_{l-1}$ and $U_{n-1} \cong V_{l-1}$. The theorem follows by induction.

It is also possible to determine when a finite inverse semigroup I is a direct product of monogenic inverse semigroups $U_1, \ldots, U_n$, with $r_i > 1$ for each i, $1 \leqslant i \leqslant n$. I must be a non-commutative, non-monogenic inverse semigroup. The direct factors can be found using similar arguments to those of section 3.

References

1961 and 1967 A.H. Clifford and G.B. Preston, *The Algebraic Theory of Semigroups*. Vols I and II, Amer. Math. Soc. Surveys 7, Providence, Rhode Island, 1961 and 1967.

TO BE PUBLISHED

[A] D.C. Trueman, *Proc. Royal Soc. of Edinburgh*, Series A, to appear.

GRAVITY DEPTH AND HOMOGENEITY IN FULL TRANSFORMATION SEMIGROUPS

J. M. Howie

Mathematical Institute, University of St. Andrews, North Haugh,
St Andrews, KY16 9SS, U.K.

Abstract

It is known that if S is the semigroup of singular mappings of a finite set $X = \{1,2,\ldots,n\}$ and E is the set of idempotents in S of rank $n - 1$ then $\langle E \rangle = S$. This result is refined by the obtaining of a formula for the least k such that $E^k = S$. Indeed, for each α in S it is possible to obtain a formula in terms of a parameter called the *gravity* of α for the least k such that $\alpha \in E^k$.

The *depth* of an idempotent-generated semigroup T with set 1 of idempotents is defined as the least k for which $I^k = T$. The depth of S is shown to be $[\frac{3}{2}(n - 1)]$. Various subsemigroups of $I(X)$ are investigated in the case where X is infinite. In this case $\langle I \rangle$ is of infinite depth, but for each infinite cardinal $\underset{\sim}{m} \leqslant |X|$ the subsemigroup $H_{\underset{\sim}{m}}$ consisting of all *homogeneous* elements of type $\underset{\sim}{m}$, i.e. all elements with shift, defect and collapse equal to $\underset{\sim}{m}$, is an idempotent-generated semigroup of depth 4.

This talk arises out of a revisiting of a paper [Howie 1966] I wrote some years ago. Let $X = \{1,2,\ldots,n\}$ be a finite set, where $n \geqslant 2$. It was proved that the semigroup $S = T(X)\setminus G(X)$ of singular mappings of X is generated by the set E of idempotents in S of rank $n - 1$. The set E has $n(n - 1)$ elements. In [Howie 1978] this result is refined, it being shown that if $n \geqslant 3$ a minimal generating set $I \subset E$ can be chosen with $|I| = \frac{1}{2}n(n - 1)$.

It is possible to refine in a different direction. Using the whole of E as a generating set, and noting that

$$E \subseteq E^2 \subseteq E^3 \subseteq \ldots,$$

we see that this ascent cannot be infinite, since S is finite. Hence

ISBN 0-12-319450-4

there exists a least $k \geqslant 1$ such that $E^k = E^{k+1} = \ldots = \langle E \rangle = S$.

It is clear that k will be a function of n, but the analysis in [Howie 1966] gives little clue as to the nature of the function. So we have:

QUESTION 1. *In terms of n, what is the least k for which $E^k = S$?*

One can also ask a related but somewhat more detailed question:

QUESTION 2. *For each α in S there is a smallest m such that $\alpha \in E^m$. Can we say what this m is in terms of reasonably accessible parameters associated with α?*

In fact we shall begin by looking at the second question and shall eventually use our result to tackle the first question. Accordingly, let α be an element of S. In a manner analogous to the splitting of a permutation into disjoint cycles we can partition X into (α-)*orbits*, the equivalence relation ω associated with this partition being given by the rule that $x \,\omega\, y$ if and only if there exist $m,n \geqslant 0$ such that $x\alpha^m = y\alpha^n$.

It is surely well-known, but hard to track down precisely in print, that each α-orbit Ω has a kernel $K(\Omega)$ characterised by the property that $x \in K(\Omega)$ if and only if $x\alpha^r = x$ for some $r > 0$. Moreover, for each y in $\Omega \setminus K(\Omega)$ there exists $s > 0$ such that $y\alpha^s \in K(\Omega)$.

If $\Omega = K(\Omega)$ and $|\Omega| > 1$ we shall say that Ω is a *cyclic* orbit. If $|\Omega| = |K(\Omega)| = 1$ we shall say that Ω is a *singleton* orbit. If $K(\Omega) \subset \Omega$ (properly) and $|K(\Omega)| = 1$ we shall say that Ω is an *acyclic* orbit.

Define $c(\alpha)$ to be the number of cyclic orbits of α and $f(\alpha)$ to be the number of fixed points. Notice that $f(\alpha)$ is the sum of the number of singleton orbits and the number of acyclic orbits.

An example may clarify matters at this stage. Let $n = 14$ and let

$$\alpha = \begin{pmatrix} 1 & 2 & 3 & 4 & 5 & 6 & 7 & 8 & 9 & 10 & 11 & 12 & 13 & 14 \\ 3 & 3 & 5 & 6 & 6 & 5 & 9 & 9 & 9 & 11 & 12 & 10 & 13 & 14 \end{pmatrix}.$$

The orbits of α can be pictured as

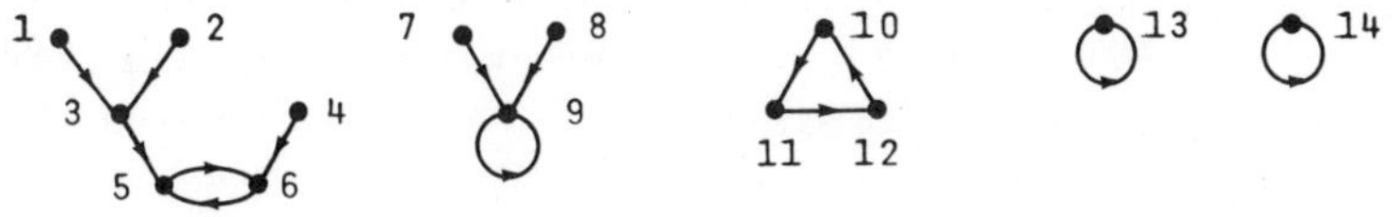

Here we have one acyclic orbit {7,8,9}, one cyclic orbit {10,11,12}, and two singleton orbits {13}, {14}. Hence

$$c(\alpha) = 1, \quad f(\alpha) = 3.$$

For each α in S we define $g(\alpha)$, the *gravity* of α, by

$$g(\alpha) = n + c(\alpha) - f(\alpha);$$

thus in our example we have $g(\alpha) = 12$. Notice that if $\alpha \in E$ then $f(\alpha) = n - 1$ and $c(\alpha) = 0$; hence $g(\alpha) = n - (n - 1) = 1$. The justification for introducing such a curious concept lies in the following result:

THEOREM 1. *Let $S = T(X)\backslash G(X)$, where $|X| = n$, and let E be the set of idempotents in S of rank $n - 1$. If $\alpha \in S$ then $\alpha \in E^{g(\alpha)}$, and $\alpha \notin E^k$ when $k < g(\alpha)$.*

The proof of this is quite technical, though the strategy is simple enough. The easier part is to show that $\alpha \in E^{g(\alpha)}$ and for the purposes of this talk it will suffice to demonstrate this for the example we have already considered. The proof in full will appear in [Howie A].

First, it is useful to recall the notation introduced in [Howie 1978] for idempotents of rank $n - 1$. For all i,j in $\{1,\ldots,n\}$ with $i \neq j$ we denote the idempotent ε for which $i\varepsilon = j$ and $x\varepsilon = x$ $(x \neq i)$ by

$$\begin{pmatrix} i \\ j \end{pmatrix}.$$

We essentially tackle α orbit by orbit (ignoring the singleton orbits, which present no problem). We express α as $\zeta\eta\tau$, where

$$\zeta = \begin{pmatrix} 6 \\ 3 \end{pmatrix}\begin{pmatrix} 5 \\ 6 \end{pmatrix}\begin{pmatrix} 3 \\ 5 \end{pmatrix}\begin{pmatrix} 1 \\ 3 \end{pmatrix}\begin{pmatrix} 2 \\ 3 \end{pmatrix}\begin{pmatrix} 4 \\ 6 \end{pmatrix},$$

$$\eta = \begin{pmatrix} 7 \\ 9 \end{pmatrix}\begin{pmatrix} 8 \\ 9 \end{pmatrix},$$

$$\tau = \begin{pmatrix} 10 \\ 8 \end{pmatrix}\begin{pmatrix} 12 \\ 10 \end{pmatrix}\begin{pmatrix} 11 \\ 12 \end{pmatrix}\begin{pmatrix} 8 \\ 11 \end{pmatrix},$$

i.e., as a product of $g(\alpha) = 12$ idempotents in E. Each orbit Ω of general form requires precisely $|\Omega|$ idempotents, each acyclic orbit requires $|\Omega| - 1$, while each cyclic orbit, because of the need for a preliminary 'linking' idempotent, requires $|\Omega| + 1$.

The proof that $g(\alpha)$ is best possible is rather harder. The essence is to prove that, for every α in S and every ε in E,

$$g(\alpha\varepsilon) \leqslant g(\alpha) + 1. \tag{1}$$

From this, and from the remark already made that elements ε of E have $g(\varepsilon) = 1$, it follows that if $\varepsilon_1, \ldots, \varepsilon_k \in E$ then

$$g(\varepsilon_1 \ldots \varepsilon_k) \leqslant k.$$

This is clearly sufficient.

The proof of formula (1) involves showing that

$$f(\alpha\varepsilon) \geqslant f(\alpha) - 1, \tag{2}$$
$$c(\alpha\varepsilon) \leqslant c(\alpha) + 1. \tag{3}$$

The first of these is easy but the second involves a detailed analysis of the way in which the orbital structure of α can be affected by post-multiplication by ε.

These inequalities of themselves would imply merely that

$$\begin{aligned} g(\alpha\varepsilon) &= n + c(\alpha\varepsilon) - f(\alpha\varepsilon) \\ &\leqslant n + (c(\alpha) + 1) - (f(\alpha) - 1) = g(\alpha) + 2, \end{aligned}$$

but in fact it is possible to demonstrate that *equality* cannot hold simultaneously in (2) and (3) for a given α and ε, and so one can deduce the stronger result (1).

Theorem 1 gives a highly satisfactory answer to question 2, and gives us moreover the means of answering question 1, since all we need do now is to identify the elements of S for which the gravity takes its greatest possible value. Clearly we maximise the gravity by maximising the number of cyclic orbits, while keeping the number of fixed points as small as possible. Equally obviously, we maximise the number of cycles by confining ourselves to 2-cycles. Thus if n is even the largest value of $c(\alpha)$ we can obtain is $\frac{1}{2}(n - 2)$. This accounts for $n - 2$ elements of X and (remembering that α must be singular) we see that the remaining two elements x,y must then either both map to x or both to y. Thus $f(\alpha) = 1$ and so

$$g(\alpha) = n + \tfrac{1}{2}(n - 2) - 1 = \frac{3}{2}n - 2.$$

Still with n even (provided $n \geqslant 4$), if we want to avoid fixed points in α we have to settle for $\tfrac{1}{2}(n - 4)$ cycles. We can then map the remaining four elements x,y,z,t in such a way as to avoid fixed points - e.g.

$$\begin{pmatrix} x & y & z & t \\ y & z & t & z \end{pmatrix}$$

- and obtain once again that

$$g(\alpha) = n + \tfrac{1}{2}(n - 4) = \frac{3}{2}n - 2.$$

If n is odd then the unique way of maximising the gravity is to have $\tfrac{1}{2}(n - 3)$ 2-cycles and then to map the remaining three elements x,y,z so as to avoid fixed points - e.g.

$$\begin{pmatrix} x & y & z \\ y & z & y \end{pmatrix}.$$

This gives

$$g(\alpha) = n + \tfrac{1}{2}(n - 3) = \frac{3}{2}(n - 1).$$

Using the notation $[r]$ for the *integral part* of a real number r we can combine the even and odd cases in a theorem as follows:

THEOREM 2. *Let S be the semigroup of singular mappings of $X = \{1,2,\dots,n\}$ and let E be the set of idempotents of S of rank $n - 1$. Let $\delta(n) = \left[\frac{3}{2}(n - 1)\right]$. Then $E^{\delta(n)} = S$, and $E^k \subset S$ for all $k < \delta(n)$.*

If T is an idempotent-generated semigroup with set I of idempotents, it is reasonable to define the *depth* $\Delta(T)$ of T as the least k for which $I^k = T$. (This will always be finite if T is finite, but if T is infinite we may have

$$T = \bigcup_{n=1}^{\infty} I^n,$$

but $I^n \subset T$ for all n; in such a case we shall define $\Delta(T) = \infty$.)

In theorem 2 we have of course made use only of some of the idempotents in S, namely those of rank $n - 1$. However, the elements of maximum gravity included some of rank $n - 1$, and in generating such elements we are forced to make exclusive use of idempotents of rank $n - 1$. Hence we can make the following deduction from theorem 2:

COROLLARY 3. $\Delta(S) = [\frac{3}{2}(n - 1)]$.

Much of [Howie 1966] was devoted to the consideration of $T(X)$ in the case where X is infinite. Continuing our theme of revisiting that paper we now let X be an infinite set of cardinal m and let E be the set of all singular idempotents of $T(X)$, i.e. the set of all idempotents except 1_X. In [Howie 1966] the subsemigroup $\langle E \rangle$ of $T(X)$ is described in terms of three cardinal numbers associated with each α in $T(X)$, called the *shift*, the *defect* and the *collapse*. The *shift* of α was the cardinal of the set

$$S(\alpha) = \{x \in X : x\alpha \neq x\},$$

the *defect* was the cardinal of

$$Z(\alpha) = X \setminus \operatorname{ran} \alpha,$$

while the *collapse* was the cardinal of

$$C(\alpha) = \cup\{t\alpha^{-1} : t \in \operatorname{ran} \alpha,\ |t\alpha^{-1}| \geqslant 2\}.$$

Let F be the set of singular mappings of finite shift, and for each infinite cardinal $p \leqslant m$ let H_p, the set of *homogeneous elements of type* p, be the set of all elements with shift, defect and collapse equal to p. Then

$$\langle E \rangle = F \cup \Big(\bigcup_{p \leqslant |X|} H_p \Big).$$

I should perhaps mention here, what has been pointed out to me privately by Professor J.N. Crossley and Professor G.B. Preston, that at one stage in the proof of this result (in the proof of Lemma 7) the argument is invalid for a *singular* cardinal. (See [Bachmann 1955] or [Preston 1962] for this definition.) However, a correct alternative argument is available and the result survives for a set X of arbitrary cardinal number.

If we examine this result more carefully we see that the proof implies that

$$\langle E \cap F \rangle = F.$$

The semigroup F is thus idempotent-generated. Less obviously, we have

THEOREM 4. *Let X be an infinite set and let F be the idempotent-generated subsemigroup of $T(X)$ consisting of all singular elements of finite shift. Then*

$$\Delta(F) = \infty.$$

This result depends on the following fairly easily verified facts:

$$(\forall\alpha,\beta \in T(X))\,|S(\alpha\beta)| \leqslant |S(\alpha)| + |S(\beta)|, \tag{4}$$

$$(\forall\alpha,\beta \in T(X))\,|Z(\alpha\beta)| \geqslant |Z(\beta)|, \tag{5}$$

$$(\forall\alpha,\beta \in F) \quad |Z(\alpha\beta)| \geqslant |Z(\alpha)|, \tag{6}$$

$$(\forall\varepsilon \in E) \qquad |Z(\varepsilon)| = |S(\varepsilon)|. \tag{7}$$

Now let $\alpha \in F$ have finite shift s and (necessarily finite) defect $d \geqslant 1$, and suppose that

$$\alpha = \varepsilon_1 \ldots \varepsilon_k \in (E \cap F)^k.$$

Then by (5) and (6) each of $\varepsilon_1,\ldots,\varepsilon_k$ must have defect at most d, and so by (7) must have shift at most d. Hence by (4) we must have $S(\alpha) \leqslant kd$. What this amounts to is that an element α is in $(E \cap F)^k$ only if

$$\frac{|S(\alpha)|}{|Z(\alpha)|} \leqslant k.$$

It follows that $\Delta(F) = \infty$ if we can produce for each $k > 0$ an element α in F for which

$$\frac{|S(\alpha)|}{|Z(\alpha)|} > k.$$

This can certainly be done: the element

$$\alpha = \begin{pmatrix} 1 & 2 & 3 & \ldots & k+1 & k+2 & \ldots \\ 2 & 3 & 4 & \ldots & k+2 & k+2 & \ldots \end{pmatrix}$$

is such an element, since $|S(\alpha)| = k + 1$, $|Z(\alpha)| = 1$.

The set H_p of homogeneous elements of type p is a subsemigroup of $T(X)$ and is in fact regular. Indeed it has the stronger property that if $\alpha \in H_p$ then every inverse of α in $T(X)$ lies in H_p. For each α in H_p we have

$$\alpha = \varepsilon_1 \ldots \varepsilon_k,$$

where $\varepsilon_1, \ldots, \varepsilon_k \in E$. In fact no ε_i may have defect greater than p and so

$$\varepsilon_1, \ldots, \varepsilon_k \in E \cap [F \cup (\bigcup_{\aleph_0 \leq q \leq p} H_q)].$$

Using a technique described in [Hall 1973] we choose an inverse α' of α and then express α as $\eta_1 \ldots \eta_k$, where, for $i = 1, \ldots, k$,

$$\eta_i = \varepsilon_i \ldots \varepsilon_k \, \alpha' \, \varepsilon_1 \ldots \varepsilon_i.$$

Then it is not too hard to see that each α_i is in $E \cap H_p$. Thus

$$H_p = \langle E \cap H_p \rangle ,$$

so that H_p is idempotent-generated.

One might expect H_p to be of infinite depth, but this is not the case. It is a consequence of the proof in [Howie 1966] that

$$H_p \subseteq E^4.$$

From this, by putting $k = 4$ in the argument above, we conclude that

$$(E \cap H_p)^4 = H_p.$$

More than that, we have

THEOREM 5. *Let X be an infinite set, let p be an infinite cardinal not greater than $|X|$ and let H_p be the idempotent-generated subsemigroup of $T(X)$ consisting of all homogeneous elements of type p. Then $\Delta(H_p) = 4$.*

To prove this, all that remains is to show that $(E \cap H_p)^3 \neq H_p$. This amounts to producing an example of an element α which cannot be expressed as a product of 3 idempotents. The technicalities are tedious in general, but a suitable α in the case where $p = |X| = \aleph_0$ is

$$\alpha = \begin{pmatrix} \{1,3,5,7,\ldots\} & 2 & 4 & 6 & 8 & \ldots \\ 2 & 1 & 3 & 5 & 7 & \ldots \end{pmatrix}.$$

The details will appear in [Howie B].

References

1955 H. Bachmann, *Transfinite Zahlen*. Ergebnisse der Math. N.F.1, Springer, 1955.

1962 G.B. Preston, A characterization of inaccessible cardinals, *Proc. Glasgow Math. Assoc.*, 5 (1962), 153-157.

1966 J.M. Howie, The subsemigroups generated by the idempotents of a full transformation semigroup. *J. London Math. Soc.*, 41 (1966), 707-716.

1973 T.E. Hall, On regular semigroups. *J. Algebra*, 24 (1973), 1-24.

1978 J.M. Howie, Idempotent generators in finite full transformation semigroups. *Proc. Royal Soc. Edinburgh*, 81A (1978), 317-323.

NOT YET PUBLISHED

[A] J.M. Howie, Products of idempotents in finite full transformation semigroups. *Proc. Royal Soc. Edinburgh A* (submitted).

[B] J.M. Howie, Some subsemigroups of infinite full transformation semigroups. *Proc. Royal Soc. Edinburgh B* (submitted).

THE SEMIGROUP OF SINGULAR ENDOMORPHISMS OF A FINITE DIMENSIONAL VECTOR SPACE

R. J. H. Dawlings

Mathematical Institute, University of St. Andrews, North Haugh, St Andrews, KY16 9SS, U.K.

Abstract

J.A. Erdös [1967b] proved that the semigroup $Sing_n$ of singular endomorphisms of an n-dimensional vector space V is generated by the set E of idempotent endomorphisms of rank $n - 1$. His proof depended entirely on matrix theory and shed very little light on the structure of the semigroup. The object of the first half of this talk is to give a more illuminating proof of this result.

From the proof given by Erdös it could be deduced that any element of $Sing_n$ could be expressed as the product of $2n$ elements of E. The second half of this talk will be devoted to reducing this bound to n, which can then be shown to be best possible.

If PF_{n-1} denotes those elements of $Sing_n$ with rank $n - 1$ and H denotes any group $\mathcal{H}$-class of $Sing_n$ contained in PF_{n-1} then I shall show that $Sing_n = \langle PF_{n-1} \rangle$ and $PF_{n-1} \subseteq \langle E \cup H \rangle$. To complete the proof that $\langle E \rangle = Sing_n$ I shall show that E generates one particular $\mathcal{H}$-class in PF_{n-1}.

So the first result is

LEMMA 1. *PF_{n-1} generates $Sing_n$.*

Proof. The proof will be by induction on the nullity of elements of $Sing_n$. Suppose, as the hypothesis, that if $\alpha \in Sing_n$ and the dimension of the null-space ${}_{\alpha}$ of α has dimension less than or equal to k then $\alpha \in \langle PF_{n-1} \rangle$. Now let $\beta \in Sing_n$ be such that $dim\ {}_{\beta} = k + 1$.

ISBN 0-12-319450-4

Let $\underset{\sim}{N}_{\beta}$ have basis $\{u_1, u_2, \ldots, u_{k+1}\}$ and extend this to a basis $\{u_1, \ldots, u_n\}$ of V. Let v be any element of V not in the range $\underset{\sim}{R}_{\beta}$ of β. Now let $\beta_1, \beta_2 \in Sing_n$ be given by

$$u_i\beta_1 = \begin{cases} u_i & i \neq k+1 \\ 0 & i = k+1 \end{cases}$$

and

$$u_i\beta_2 = \begin{cases} u_i\beta & i \neq k+1 \\ v & i = k+1 \end{cases}$$

It is easy to check that $\beta = \beta_1\beta_2$. Now $\beta_1 \in PF_{n-1}$ and $dim\ \underset{\sim}{R}_{\beta_2} = dim\ \underset{\sim}{R}_{\beta} + 1$, i.e. $dim\ \underset{\sim}{N}_{\beta_2} = k$. Thus $\beta_1, \beta_2 \in \langle PF_{n-1} \rangle$ and consequently $\beta \in \langle PF_{n-1} \rangle$. The induction process may be started since any element with nullity 1 belongs to PF_{n-1}.

To show that $PF_{n-1} \subseteq \langle E \cup H \rangle$ the following lemma, whose proof is omitted, will be used.

LEMMA 2. *If ϕ, γ are idempotents in PF_{n-1} then there exists an idempotent ε in PF_{n-1} such that $\phi\ \varepsilon\ \gamma \in PF_{n-1}$.*

We now have

LEMMA 3. *Let H be any H-class of $Sing_n$ contained in PF_{n-1} and let E be the set of idempotents in PF_{n-1}. Then $E \cup H$ generates PF_{n-1}.*

Proof. Let H' be any H-class of $Sing_n$ in the same R-class as H. Since $Sing_n$ is regular there exist elements ϕ and γ of E such that ϕ is in the same L-class as H and γ is in the same L-class as H'. By lemma 2 there exists an $\varepsilon \in E$ such that $\phi\ \varepsilon\ \gamma \in PF_{n-1}$.

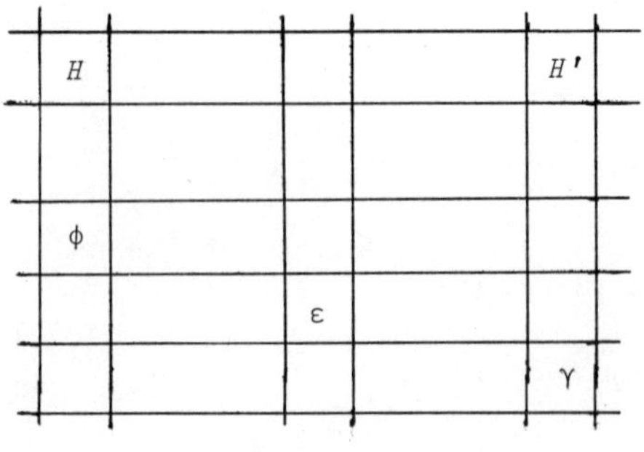

Let $\alpha \in H$. Now, since I am writing my mappings on the right,

$$\underset{\sim}{R}_{\alpha\varepsilon\gamma} \subseteq \underset{\sim}{R}_{\gamma} \text{ and } \underset{\sim}{N}_{\alpha\varepsilon\gamma} \supseteq \underset{\sim}{N}_{\alpha}.$$

But $\alpha\phi = \alpha$ and so $\alpha\ \varepsilon\ \gamma = (\alpha\phi)(\phi\ \varepsilon\ \gamma)$.

Now $\alpha\phi \neq 0$ in PF^0_{n-1} and $\phi\ \varepsilon\ \gamma \neq 0$ in PF^0_{n-1}. Hence since the semigroup PF^0_{n-1} is categorical at zero (see [1967a exercise 7.7.5]) it follows that $\alpha\ \varepsilon\ \gamma \neq 0$ in PF^0_{n-1}, i.e. that $\alpha\ \varepsilon\ \gamma$ has rank $n - 1$. From this it now follows that

$$\underset{\sim}{R}_{\alpha\varepsilon\gamma} = \underset{\sim}{R}_{\gamma} \text{ and } \underset{\sim}{N}_{\alpha\varepsilon\gamma} = \underset{\sim}{N}_{\alpha}.$$

But two elements are $\mathcal{L}$-($\mathcal{R}$-) equivalent if and only if they have the same range (null-space) ([1961, exercise 2.2.6]) and so $\alpha\ \varepsilon\ \gamma\ \mathcal{L}\ \gamma$ and $\alpha\ \varepsilon\ \gamma\ \mathcal{R}\ \alpha$. But since α is in the same $\mathcal{R}$-class as H' and γ is in the same $\mathcal{L}$-class as H' we thus have $\alpha\ \varepsilon\ \gamma \in H'$. In fact the map from H to H' given by $\alpha \to \alpha\ \varepsilon\ \gamma$ is a bijection by Greens Lemma ([1961, lemma 2.2]). Thus any element of H' is a product of elements of H and E. Hence the $\mathcal{R}$-class containing H is generated by $H \cup E$.

Now consider any $\mathcal{H}$-class H'' in the same $\mathcal{D}$-class as H. Then there exists an $\mathcal{H}$-class H' in the same $\mathcal{L}$-class as H'' and the same $\mathcal{R}$-class as H. Again since $Sing_n$ is regular there exist elements ϕ_1, γ_1 of E such that ϕ_1 is in the same $\mathcal{R}$-class as H' and γ_1 is in the same $\mathcal{R}$-class as H''. By lemma 2 there exists an element ε_1 in E such that $\gamma_1\ \varepsilon_1\ \phi_1 \in PF_{n-1}$.

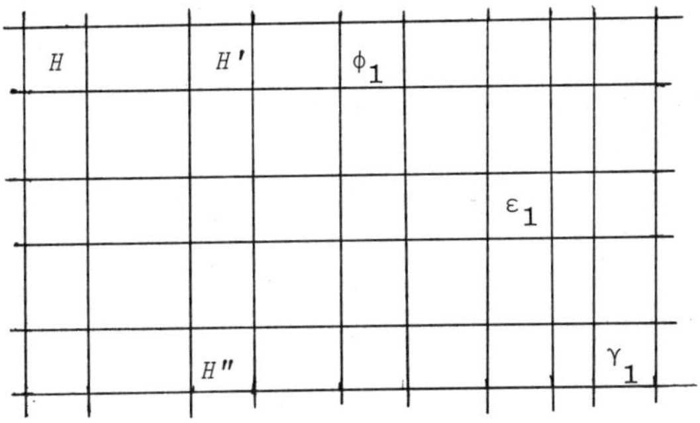

Dual to the above, the map from H' to H'' given by $\alpha \to \gamma_1\ \varepsilon_1\ \alpha$ $(\forall\alpha \in H')$ is a bijection. Thus H'' is generated by $H' \cup E$. But H' is generated by $H \cup E$ and so H'' is generated by $H \cup E$. Since H'' was an arbitrary $\mathcal{H}$-class of the $\mathcal{D}$-class containing H we have that the $\mathcal{D}$-class containing

H is generated by $H \cup E$. But two elements of $Sing_n$ are $\mathcal{D}$-equivalent if and only if they have the same rank ([1961, exercise 2.2.6]). Thus PF_{n-1} is generated by $E \cup H$.

All that remains now is to show that E generates some $\mathcal{H}$-class of $Sing_n$ contained in PF_{n-1}.

LEMMA 4. *Let V be an n-dimensional vector space over the field F. Let H be the $\mathcal{H}$-class of $Sing_n$ containing those elements with null-space $\{(a,0,0,0,\ldots,0): a \in F\}$ and range $\{(a_1,0,a_3,a_4,\ldots,a_n): a_i \in F\}$. Then E generates H.*

Proof. The proof is by induction on n. Suppose the lemma is true for dimension $n - 1$. Then, by lemmas 1 and 3, the idempotent endomorphisms of rank $n - 2$ generate $Sing_{n-1}$.

Now let $\alpha \in H$. Then α has matrix of the form

$$M = \left[\begin{array}{c|c|c} 0 & 0 & 0 \\ \hline a & 0 & M' \end{array}\right].$$

Now $M = A_1B$ where

$$A_1 = \left[\begin{array}{c|c} 0 & 0 \\ \hline a & I_{n-1} \end{array}\right],$$

$$B = \left[\begin{array}{c|c|c} 1 & 0 & 0 \\ \hline 0 & 0 & M' \end{array}\right]$$

and I_i is the $i \times i$ identity matrix. Notice that A_1 is idempotent. Now

$$B' = \left[\begin{array}{c|c} 0 & M' \end{array}\right]$$

represents an element of rank $n - 2$ of $Sing_{n-1}$. So, by the hypothesis,

$$B' = A'_2 A'_3 \ldots A'_m$$

where $A'_2, A'_3, \ldots, A'_m$ are idempotent $(n - 1) \times (n - 1)$ matrices of rank $n - 2$. Hence

$$B = A_2 A_3 \ldots A_m,$$

where

$$A_i = \left[\begin{array}{c|c} 1 & 0 \\ \hline 0 & A'_i \\ & \end{array}\right], \quad (i = 2, 3, \ldots, m).$$

Notice that each A_i $(i = 2, 3, \ldots, m)$ is idempotent. So

$$M = A_1 A_2 \ldots A_m$$

where each A_i $(i = 1, 2, \ldots, m)$ is idempotent.

To anchor the induction notice that the result is trivial for $n = 1$ (for $Sing_1 = \{[0]\}$, the set consisting of the zero matrix).

So we now have

THEOREM 1. *Let $Sing_n$ denote the semigroup of singular endomorphisms of an n-dimensional vector space and let E denote the idempotent elements of $Sing_n$ with rank $n - 1$. Then E generates $Sing_n$.*

If the field over which the vector space is defined is finite then the semigroup $Sing_n$ is finite. So the chain

$$E \subseteq E^2 \subseteq E^3 \subseteq E^4 \subseteq \ldots$$

must at some point become stationary. Since E generates $Sing_n$ there must be an integer m such that

$$E^m = E^{m+1} = \ldots = Sing_n.$$

So it is reasonable to try to find the least integer m satisfying $E^m = Sing_n$.

If the field is infinite we must first ask whether an integer m exists such that $E^m = Sing_n$. If an m does exist we may then try to find its minimum value.

From the proof of lemma 4 it is tempting to guess that $E^n = Sing_n$. This result is, in fact, true even though the H-class considered in lemma 4 was not a general H-class. As the proof of this is technical I shall only give a rough outline. Before giving this outline proof I shall state two lemmas which will be needed. The proofs of these, also technical, are omitted.

LEMMA 5. *Let A be the $k \times k$ matrix*

$$\begin{bmatrix} 0 & 1 & 0 & 0 & 0 & \dots & 0 \\ 0 & 0 & 1 & 0 & 0 & \dots & 0 \\ 0 & 0 & 0 & 1 & 0 & \dots & 0 \\ \cdot & & & & \cdot & & \\ \cdot & & & & & \cdot & \\ \cdot & & & & & & \cdot \\ 0 & 0 & 0 & 0 & 0 & \dots & 1 \\ 0 & a_2 & a_3 & a_4 & a_5 & \dots & a_k \end{bmatrix}.$$

Then $A = B\, E_0\, E_1 \dots E_{k-2}$ where

$$B = \begin{bmatrix} 1 & 0 & 0 & 0 & \dots & 0 & 0 \\ 0 & 1 & 0 & 0 & \dots & 0 & 0 \\ 0 & 0 & 1 & 0 & \dots & 0 & 0 \\ \cdot & & \cdot & & & & \\ \cdot & & & \cdot & & & \\ \cdot & & & & \cdot & & \\ 0 & 0 & 0 & 0 & \dots & 1 & 0 \\ a_2 & a_3 & a_4 & a_5 & \dots & a_k & 0 \end{bmatrix},$$

$$E_i = \left[\begin{array}{c|cc|c} I_{k-2-i} & 0 & & 0 \\ \hline 0 & \begin{matrix} 0 \\ 0 \end{matrix} & \begin{matrix} 1 \\ 1 \end{matrix} & 0 \\ \hline 0 & 0 & & I_i \end{array}\right]$$

and I_j is the $j \times j$ identity matrix. Notice that B and each E_i are idempotent and have nullity 1.

LEMMA 6. *Let C be the $(k + 1) \times (k + 1)$ matrix*

$$\left[\begin{array}{cccccc|c} 0 & 1 & 0 & 0 & 0 \ldots & 0 & \\ 0 & 0 & 1 & 0 & 0 \ldots & 0 & \\ 0 & 0 & 0 & 1 & 0 \ldots & 0 & \\ \cdot & & & & \cdot & & \\ \cdot & & & & & \cdot & 0 \\ \cdot & & & & & \cdot & \\ 0 & 0 & 0 & 0 & 0 \ldots & 1 & \\ a_1 & a_2 & a_3 & a_4 & a_5 \ldots & a_k & \\ \hline & & & 0 & & & 0 \end{array}\right]$$

Then $C = D F_1 F_2 \ldots F_{k-1} G$ *where*

$$D = \left[\begin{array}{c|cc} I_{k-1} & & 0 \\ \hline 0 & 1 & 1 \\ & 0 & 0 \end{array}\right], \qquad F_i = \left[\begin{array}{c|cc|c} I_{k-1-i} & & 0 & 0 \\ \hline & 0 & 1 & \\ 0 & 0 & 1 & 0 \\ \hline 0 & & 0 & I_i \end{array}\right],$$

$$G = \left[\begin{array}{ccccc|c} & & I_k & & & 0 \\ \hline a_1 & a_2 & \ldots & a_{k-1} & a_k - 1 & 0 \end{array}\right]$$

and I_j *is the* $j \times j$ *identity matrix. Notice that* D, G *and each* F_i *are idempotent and have nullity 1.*

THEOREM 2. *Let* $Sing_n$ *denote the semigroup of singular endomorphisms of an n-dimensional vector space and let E be the set of idempotents of* $Sing_n$ *of rank* $n - 1$. *Then* $E^n = Sing_n$.

Sketch Proof. We need only consider matrices of the form $diag\ \{A_1, A_2, \ldots, A_p\}$ where each A_i is an $r_i \times r_i$ matrix of the form

$$A_i = \left[\begin{array}{cccccc} 0 & 1 & 0 & 0 & 0 \ldots & 0 \\ 0 & 0 & 1 & 0 & 0 \ldots & 0 \\ 0 & 0 & 0 & 1 & 0 \ldots & 0 \\ \cdot & & & & \cdot & \\ \cdot & & & & & \cdot \\ \cdot & & & & & \cdot \\ 0 & 0 & 0 & 0 & 0 \ldots & 1 \\ a_{i1} & a_{i2} & a_{i3} & a_{i4} & a_{i5} \ldots & a_{ir_i} \end{array}\right]$$

since every matrix is similar to one of this form. (This is the rational canonical form of a matrix. See, for example, [1974].) Since we are only concerned with singular matrices we may assume, without loss of generality, that A_p is singular, i.e. that $a_{p1} = 0$. So, by lemma 5,

$$A_p = J_1 J_2 \dots J_{r_p},$$

where each J_i is idempotent with nullity 1. Thus

$$A = A^{(1)} J_2' J_3' \dots J_{r_p}'$$

where

$$A^{(1)} = \begin{bmatrix} A_1 & & & & \\ & A_2 & & 0 & \\ & & \ddots & & \\ & 0 & & A_{p-1} & \\ & & & & J_1 \end{bmatrix} \text{ and } J_i' = \left[\begin{array}{c|c} I_{n-r_p} & 0 \\ \hline 0 & J_i \end{array}\right] \quad (i = 2,3,\dots,r_p).$$

Thus A is the product of $A^{(1)}$ with $r_p - 1$ elements of E.

Now consider the submatrix

$$J_1' = \begin{bmatrix} A_{p-1} & 0 \\ 0 & J_1 \end{bmatrix}$$

of $A^{(1)}$. Since J_1 is an idempotent $r_p \times r_p$ matrix with nullity 1, J_1 is similar to

$$\left[\begin{array}{c|c} 0 & 0 \\ \hline 0 & I_{r_p-1} \end{array}\right].$$

Thus J_1' is similar to

$$J_1'' = \left[\begin{array}{c|c|c} A_{p-1} & 0 & 0 \\ \hline 0 & 0 & 0 \\ \hline 0 & 0 & I_{r_p-1} \end{array}\right].$$

Now consider the submatrix

$$A'_{p-1} = \left[\begin{array}{c|c} A_{p-1} & 0 \\ \hline 0 & 0 \end{array}\right]$$

of J''_1. By lemma 6, A'_{p-1} is a product of $r_{p-1} + 1$ idempotents with nullity 1. Thus J''_1 is a product of $r_{p-1} + 1$ idempotents with nullity 1. But J'_1 is similar to J''_1, so

$$J'_1 = K'_1 K'_2 \ldots K'_{r_{p-1}+1}$$

where each K'_i is an $(r_p + r_{p-1}) \times (r_p + r_{p-1})$ idempotent matrix with nullity 1. Consequently

$$A^{(1)} = A^{(2)} K_2 K_3 \ldots K_{r_{p-1}+1}$$

where

$$A^{(2)} = \begin{bmatrix} A_1 & & & & \\ & A_2 & & 0 & \\ & & \ddots & & \\ & 0 & & A_{p-2} & \\ & & & & K'_1 \end{bmatrix} \quad \text{and} \quad K_i = \left[\begin{array}{c|c} I_{n-r_p-r_{p-1}} & 0 \\ \hline 0 & K'_i \end{array}\right]$$

$(i = 2,3,\ldots,r_{p-1} + 1)$. Thus $A^{(1)}$ is the product of $A^{(2)}$ with r_{p-1} elements of E. Hence A is the product of $A^{(2)}$ with $r_p + r_{p-1} - 1$ elements of E.

Now consider the submatrix

$$\left[\begin{array}{c|c} A_{p-2} & 0 \\ \hline 0 & K'_1 \end{array}\right]$$

and proceed in the same manner as above. Eventually we shall obtain that A is the product of $A^{(p)}$ with $\sum_{i=1}^{p} r_i - 1$ elements of E. But $A^{(p)} \in E$ and $\sum_{i=1}^{p} r_i = n$ and so we have that A is the product of n elements of E.

To show that n is the least integer k such that $E^k = Sing_n$ the following theorem, which is also interesting in its own right, is needed.

THEOREM 3. *Let $Sing_n$ denote the semigroup of singular endomorphisms of an n-dimensional vector space V and let E denote the idempotent elements of $Sing_n$ of rank $n - 1$. Let $\alpha \in Sing_n$. Then $\alpha \in E^{k(\alpha)}$ where $k(\alpha) = n - \dim\{x \in V: x\alpha = x\}$. Also, if $\ell < k(\alpha)$ then $\alpha \notin E^{\ell}$.*

Proof. Suppose $D_\alpha = \{x \in V: x\alpha = x\}$ has dimension d. Let $\{u_1, u_2, \ldots, u_d\}$ be a basis for D_α and extend this to a basis $\{u_1, \ldots, u_n\}$ for V. Relative to this basis α has matrix

$$M_\alpha = \left[\begin{array}{c|c} I_d & 0 \\ \hline 0 & M \end{array}\right]$$

where I_d is the $d \times d$ identity matrix and M is a singular $(n - d) \times (n - d)$ matrix.

By theorem 2, $M = M_1 M_2 \ldots M_{n-d}$ where M_i is idempotent of rank $n - d - 1$ $(i = 1,2,\ldots,n-d)$. Thus $M_\alpha = M_1' M_2' \ldots M_{n-d}'$ where

$$M_i' = \left[\begin{array}{c|c} I_d & 0 \\ \hline 0 & M_i \end{array}\right] \qquad (i = 1,2,\ldots,n-d).$$

Since each M_i' is idempotent of rank $n - 1$ we have that M may be expressed as the product of $n - d$ idempotent matrices of rank $n - 1$, i.e. $\alpha \in E^{n-d} = E^{k(\alpha)}$.

The proof that $k(\alpha)$ is the least possible k such that $\alpha \in E^k$ is again technical and so will be omitted. (The proof involves showing that the assumption $\alpha \in E^{k(\alpha)-1}$ leads to the contradiction of $k(\alpha) - 1$ vectors spanning a $k(\alpha)$-dimensional space.)

We are now in a position to prove the final theorem.

THEOREM 4. *Let $Sing_n$ denote the semigroup of singular endomorphisms of an n-dimensional vector space V and let E denote the idempotent elements of $Sing_n$ of rank $n - 1$. If $k < n$ then $E^k \subset Sing_n$.*

Proof. Clearly $E^k \subseteq Sing_n$. To show that $E^k \neq Sing_n$ if $k < n$ it will suffice to find an element α of $Sing_n$ such that $k(\alpha) = n$, i.e. such that $\dim\{x \in V: x\alpha = x\} = 0$. This is equivalent to finding an $n \times n$ singular matrix that does not have 1 as an eigenvalue. There is, of course, no trouble in doing this.

References

1961 A.H. Clifford and G.B. Preston, *The algebraic theory of semigroups*, Vol.I. Math. Surveys of the American Math. Soc. No.7 (1961).

1967a A.H. Clifford and G.B. Preston, *The algebraic theory of semigroups*, Vol.II. Math. Surveys of the American Math. Soc. No.7 (1967).

1967b J.A. Erdös, On products of idempotent matrices. *Glasgow Math. J.* 8(2), (1967), pp.118-122.

1974 N. Jacobson, *Basic algebra* Vol.I. W.H. Freeman and Co. (1974).

PROJECTIVES IN SOME CATEGORIES OF SEMIGROUPS

P. G. Trotter

Department of Mathematics, University of Tasmania, Hobart, Tasmania, Australia.

Abstract

Some authors define projectives in categories of algebraic systems in terms of surjective morphisms rather than epimorphisms. Differences arising from these definitions are observed here in the categories of semigroups and of finite semigroups. The definitions coincide in the categories of finite regular semigroups and of inverse semigroups; a characterization of the projectives in these categories is given.

1. Introduction

Although projectives have been studied in many categories, very little seems to be known about them in categories of semigroups. The only papers to appear on this topic, as far as the author is aware, are by Horn and Kimura [1971] and Nordahl and Scheiblich [B]; these are both concerned with projectives in categories of bands. Inverse semigroup projectives have been characterized [Trotter, C]. In work not yet submitted for publication H.E. Scheiblich has studied * band projectives and in these proceedings P.R. Jones [A] describes completely simple semigroup projectives.

Throughout this paper C will denote a category of semigroups, the morphisms of C being semigroup homomorphisms. A morphism $\beta: T \to S$ in C is an *epimorphism* if for any morphisms $\delta_1, \delta_2: S \to U$ in C, $\beta\delta_1 = \beta\delta_2$ implies $\delta_1 = \delta_2$. An object R in C is a C-*projective* if for every morphism $\alpha: R \to S$ and every epimorphism $\beta: T \to S$, there exists a morphism $\gamma: R \to T$ so that $\gamma\beta = \alpha$. We will call R a *weak* C-*projective* if β is required only to be a surjective morphism in the definition.

ISBN 0-12-319450-4

Notice that a surjective morphism is an epimorphism. The converse is false, as can be seen from the following examples due respectively to J.M. Howie and T.E. Hall.

EXAMPLES 1.1. (a) Let $\mathcal{S}$ denote the category of semigroups. Let S and T denote respectively the infinite cyclic group and cyclic subsemigroup generated by a. Define $\beta\colon T \to S$ by $x\beta = x$. It can be readily checked that β is a non-surjective epimorphism.

(b) Let $\mathcal{F}$ denote the category of finite semigroups. In this case let $S = \{e,f,g,h,0\}$ denote a five element completely 0-simple semigroup with one non-idempotent element h; the inverse of h being g. Let $T = \{e,f,h,0\}$, which is a subsemigroup of S. Again the map $\beta\colon T \to S$ given by $x\beta = x$ can be easily shown to be a non-surjective epimorphism.

In algebraic studies the notion of weak projective seems more natural than that of projective; non-surjective epimorphisms are rarely considered. In the works mentioned above, on projectives in categories of semigroups, the projectives determined are either weak projectives or are in categories where epimorphisms are all surjective. However the class of $\mathcal{C}$-projectives and the class of weak $\mathcal{C}$-projectives can differ considerably. This can be easily demonstrated using example 1.1 and a modification of the following argument due to T.E. Hall.

THEOREM 1.2. *Let $\mathcal{R}$ be the category of regular semigroups and suppose $\beta\colon T \to S$ is a non-surjective epimorphism in $\mathcal{R}$. Then there are no $\mathcal{R}$-projectives.*

Proof. By [Hall 1978; theorem 8] an amalgam $(S,S';T\beta)$, where $S \simeq S'$ and $T\beta = S \cap S'$, is embeddable in a regular semigroup if $T\beta$ contains all the idempotents of S and S'. It follows since β is a non-surjective epimorphism that $E_S \neq E_{T\beta}$, where E_S denotes the set of idempotents of S. So there exists $e = e^2 \in S \setminus T\beta$. Let $\alpha\colon R \to S$ be a morphism with $R\alpha = \{e\}$. Then there exists no $\gamma\colon R \to T$ so that $\gamma\beta = \alpha$, and hence R is not an $\mathcal{R}$-projective.

The last two sentences applied to examples 1.1 show the first part of the following:

THEOREM 1.3. *There are no S- or F-projectives. The weak S-projectives are the free semigroups. Finite chains of idempotents are weak F-projectives.*

Proof. The second statement is from [Trotter, C]. Suppose $R = \{e_1,\ldots,e_n\}$ is a chain of idempotents where $e_i > e_j$ if $i < j$. Let $\alpha\colon R \to S$ be a morphism and $\beta\colon T \to S$ be a surjective morphism in F. Let $T_1 = (e_1\alpha)\beta^{-1}$ be the set of preimages of $e_1\alpha$ under β. T_1 is a finite subsemigroup of T so there exists an idempotent $f_1 \in T_1$, and $f_1\beta = e_1\alpha$. Note that $(f_1Tf_1)\beta = (e_1\alpha)S(e_1\alpha)$ and $e_2\alpha \in (e_1\alpha)S(e_1\alpha)$. Let $\beta_1\colon f_1T_1f_1 \to (e_1\alpha)S(e_1\alpha)$ be the surjective morphism in F obtained by restricting the domain of β. Let $T_2 = (e_2\alpha)\beta_1^{-1}$. We can choose $f_2 = f_2^2 \in T_2 \leqslant f_1Tf_1$, as above, so that $f_2 \leqslant f_1$ and $f_2\beta = e_2\alpha$. Repeat the process to get an idempotent chain $f_1 \geqslant \ldots \geqslant f_n$ in T so that $f_i\beta = e_i\alpha$. Then $\gamma\colon R \to T$ given by $e_i\gamma = f_i$ is a morphism so that $\gamma\beta = \alpha$.

Some properties of C-projectives can be deduced easily from the definition. In C, S is a *retract* of T if there exists an epimorphism $\beta\colon T \to S$ and a morphism $\gamma\colon S \to T$ so that $\gamma\beta$ is the identity on S. A C-projective R is an *absolute quotient retract*; that is, R is a retract of every object T in C of which it is an epimorphic image. If epimorphisms in C are surjective and there exist free objects in C then the C-projectives are precisely the retracts of free objects. Any retract of a C-projective is a C-projective.

The author is indebted to H. Lausch for the following proof.

THEOREM 1.4. *In the category of finite groups the only projective is the trivial group.*

Proof. Note that, by chapter XIII.2 of [Maclane and Birkhoff 1967] G is a split extension of a group N by a group B if and only if B is a retract of G with respect to any epimorphism $G \to B$ with kernel N. The two results that follow are from [Lang 1966]. The extensions of an elementary abelian p-group A by a p-group P split if and only if A is a free $Z_p(P)$-module (proposition III, 2). For a finite group G with a p-Sylow subgroup P there exists a finite abelian group $M(G,P,A)$ so that the extensions of A by P split if and only if the extensions of $M(G,P,A)$ by G split (theorem II, 4). Hence given $G \neq \{1\}$ and P, choose A to be an elementary abelian p-group that is not a free

$Z_p(P)$-module. G is then not a retract of some extension of $M(G,P,A)$ by G. Notice that by corollary 15.2 of [Neumann 1954] epimorphisms are surjections in the category of finite groups. The result now follows since $G \neq \{1\}$ is not an absolute quotient retract, while $\{1\}$ is clearly a projective.

2. Finite regular semigroups

Let F_R denote the category of finite regular semigroups. The following is an unpublished result due to T.E. Hall.

THEOREM 2.1. *Epimorphisms are surjective in* F_R.

Note that each semigroup in F_R is completely semisimple. In this section let R always denote an F_R-projective with principal series $R = R_0 \supset R_1 \supset \dots \supset R_n$. Such an R exists; the one element semigroup is an F_R-projective. Define $\underline{R}_i = R_i \setminus R_{i+1}$ if $i < n$ and $\underline{R}_n = R_n$. We regard $\underline{R}_i$ as a subset of R_i/R_{i+1}.

The aim in this section is to present a procedure by which any F_R-projective can be constructed. An outline only of the justification of the procedure is provided here. A detailed version will appear elsewhere. We first observe some properties of the F_R-projective R.

THEOREM 2.2. *An* F_R*-projective* R *is a band.*

Proof. The result can be shown by an induction argument on the principal series for R listed above. The aim of the proof is to show that $R \setminus R_r$ is a band for $0 < r \leqslant n$, and that R therefore is a band. Included here is the proof that $\underline{R}_0$ is a band. Although the underlying ideas are similar in the cases where $r > 0$, the proof in these cases is complicated by the ideal extensions involved. We omit the cases where $r > 0$ from this description.

Let $R/R_1 \simeq M^0(G;I,\Lambda;P)$, the Rees matrix semigroup over the group with zero G^0 with $\Lambda \times I$ sandwich matrix $P = (p_{\lambda i})$, and with non-zero elements $(a)_{i\lambda}$, $a \in G$, $(i,\lambda) \in I \times \Lambda$. Suppose F is a finite group and $\delta: F \to G$ is an epimorphism. Put $0\delta = 0$. Let $T = M^0(F;I,\Lambda;Q)$ where $Q = (q_{\lambda i})$ and $q_{\lambda i}\delta = p_{\lambda i}$. The map $\beta: T \to R/R_1$ given by $(a)_{i\lambda}\beta = (a\delta)_{i\lambda}$ is clearly an epimorphism. Let $\alpha: R \to R/R_1$ be the natural homomorphism. There is a morphism $\gamma: R \to T$ so that $\gamma\beta = \alpha$, since R is an

F_R-projective. Let $H \simeq G$ be a group H-class of R in $\underline{R}_0$, then there is an H-class $K \simeq F$ in T so that $H\gamma \subseteq K$. But F is an arbitrary finite group epimorphic preimage of G so G is an absolute quotient retract in the category of finite groups. By theorem 1.4, $G = \{e\}$.

Now let $T_1 = M(F;I,\Lambda;Q)$ be a completely simple semigroup where $F = \langle a; a^m = 1 \rangle$ (a finite cyclic group) and $Q = (q_{\lambda i})$ is a $\Lambda \times I$ matrix with

$$q_{\lambda i} = \begin{cases} 1 & \text{if } p_{\lambda i} = 1 \\ a & \text{if } p_{\lambda i} = 0 . \end{cases}$$

Define $T = T_1 \cup \underline{R}_0$ and define a multiplication on T compatible with the multiplication on $\underline{R}_0$ and T_1 so that in T

$$\left.\begin{aligned} (e)_{i\lambda}(b)_{j\tau} &= (q_{\lambda j}b)_{i\tau} \\ (b)_{j\tau}(e)_{i\lambda} &= (bq_{\tau i})_{j\lambda} \end{aligned}\right\} \quad \text{if } (e)_{i\lambda} \in \underline{R}_0 \text{ and } (b)_{j\tau} \in T_1, \text{ and}$$

$$(e)_{i\lambda}(e)_{h\mu} = (a)_{i\mu} \qquad \text{if } (e)_{i\lambda}(e)_{h\mu} = 0 \text{ in } R/R_1.$$

It can be readily checked that $T \in F_R$; for example, see lemma III.4.1 of [Petrich 1973a]. Let $\beta\colon T \to R/R_1$ be the epimorphism that acts as the identity map on $\underline{R}_0$ and the zero map on T_1. Let $\alpha\colon R \to R/R_1$ be the natural morphism and γ be the morphism induced by the projective property. If R/R_1 has proper zero divisors then $(a)_{i\mu} \in R\gamma$ for some $(i,\mu) \in I \times \Lambda$. But $(a)_{i\mu}$ generates a subgroup of T of order m if $q_{\mu i} = 1$ or m is odd, or of order $m/2$ if $q_{\mu i} = a$ and m is even. This argument applies for all m, which contradicts the finiteness of R. Hence $\underline{R}_0$ is a rectangular band so R/R_1 is a band.

Having shown R/R_r is a band with no proper divisors of zero for all r, $0 < r \leqslant n$, and that R_n is a band, then

COROLLARY 2.3. *R is a chain of rectangular bands.*

In the next result we begin to determine the relationships between the rectangular band components of R.

LEMMA 2.4. *Suppose $a,b \in \underline{R}_i$ and $c \in R_i$ where $a \mathcal{R} b$ and $a \neq b$. Then (i) $ca \neq cb$. Suppose also that $a,b \notin \underline{R}_i(R\backslash R_i)$, $d,e \in R\backslash R_i$ and $d \neq e$. Then (ii) $ad \neq bd$ and (iii) $ad \neq ae$.*

Proof. The results are consequences of the projective definition. For (i), let $\delta: \underline{R}_i \to T_1$ be an isomorphism and define $T = T_1 \cup R\backslash R_{i+1}$ where $x(a\delta) = (xa)\delta$ and $(a\delta)x = (ax)\delta$ for $x \in R\backslash R_{i+1}$, $a \in \underline{R}_i$. Let $S = R/R_{i+1}$ and $\beta: T \to S$ be the identity map on $R\backslash R_{i+1}$ and the zero map on T_1. For (ii) and (iii) let T be a finite free band and $S = R$. The proof of claim 3 of [Nordahl and Scheiblich, B] can be readily modified to give the results.

LEMMA 2.5. *Any completely [0-]simple homomorphic image of R is a right or left zero semigroup [with adjoined zero]. In particular $\underline{R}_0$ is a right or left zero semigroup.*

Proof. This is again a consequence of the projective definition. Let S be a completely simple homomorphic image of R, so by theorem 2.2, S is an $I \times \Lambda$ rectangular band, for some sets I and Λ. Choose $T \simeq M(G;I,\Lambda,P)$ for some G and P. If $|I| \neq 1 \neq |\Lambda|$ then G and P may be chosen so that T is a non-orthodox semigroup. The result follows. If S is a completely 0-simple homomorphic image of R, then by corollary 2.3, S has no proper zero divisors. Let $T = (M(G;I,\Lambda;P))^0$ and argue as above.

COROLLARY 2.6. *Either $\underline{R}_0 R = R$ or $R\underline{R}_0 = R$. If $\underline{R}_0$ is a non-trivial right zero semigroup then $\underline{R}_0 R = R$.*

Proof. Let τ be an equivalence relation with classes $A = \underline{R}_0 R\underline{R}_0$, $B = \underline{R}_0 R\backslash A$, $C = R\underline{R}_0\backslash A$ and $D = R\backslash(A \cup B \cup C)$. It can be checked, using lemma 2.4(i), that τ is a congruence and R/τ is a rectangular band. The first statement then follows from lemma 2.5. A slightly more complex version of the argument gives the second statement.

The next result enables us to determine some additional F_R-projectives from an F_R-projective R. The proof is straightforward.

THEOREM 2.7. *Let $\sigma = \{(a,b) \in R \times R;\ ae = be \text{ for } e \in \underline{R}_0\}$. Suppose R is an F_R-projective and $\underline{R}_0$ is a right zero semigroup. Then σ is a congruence and the following are F_R-projectives: (i) R/σ, (ii) $R\backslash\{1\}$ whenever $R = R^1$ and (iii) $R\backslash R_i$, $1 \leqslant i \leqslant n$.*

The following technical result details further restrictions on the rectangular band components of the chain R. For S a rectangular band let $|S|_r$ denote the cardinality of any $\mathcal{R}$-class of S. The inequalities of the result come from lemma 2.4. The remaining statements can be obtained by repeated use of theorem 2.7 (giving subsemigroup retracts of R), lemma 2.5 and corollary 2.6.

LEMMA 2.8. *Suppose* $|\underline{R}_i \backslash \underline{R}_i(R \backslash R_i)|_r = m > 0$ *and* $|\underline{R}_i e_j \backslash \underline{R}_i(R_{j+1} \backslash R_i|_r = m_j$ *for* $0 \leqslant j < i$ *and* $e_j \in \underline{R}_j$. *Then* $m \leqslant m_0 \leqslant \ldots \leqslant m_{i-1}$, $\underline{R}_j$ *is a right zero semigroup and* $\underline{R}_j R_j = R_j$. *If* $\underline{R}_i R_i \neq R_i$ *or* $\underline{R}_i$ *is not a right zero semigroup then* $m_{i-1} = 1$ *and* $R_i \underline{R}_i = R_i$.

Theorem 2.7 indicates how new F_R-projectives can be obtained by reducing a given F_R-projective. The three lemmas that follow detail a procedure by which F_R-projectives can be constructed from an F_R-projective R by extension. Lemmas 2.10 and 2.11 together reverse the reduction of theorem 2.7(i). Their proofs are technical and their outlines are omitted here. However, at the end of this section we include a detailed description of the F_R-projectives that are chains of one or two rectangular bands. The explanations included with the description reflect the major parts of the proofs. The next result is obvious.

LEMMA 2.9. R^1 *is an* F_R-*projective.*

Suppose $R = R^1$. Let $K = \{e_1 = 1, e_2, \ldots, e_r\}$ be a right zero semigroup. Define $U(R,r) = R \times K$, the direct product of the bands R and K.

LEMMA 2.10. *If* $R = R^1$ *then* $U(R,r)$ *is an* F_R-*projective.*

With lemma 2.8 in mind, we construct the following semigroup. Assume $e \in \underline{R}_0$, $\underline{R}_j$ is a right zero semigroup, $\underline{R}_j R_j = R_j$ for $0 \leqslant j < i$, and $\underline{R}_i$ is an $I \times \Lambda$ rectangular subband of an $I \times \Omega$ rectangular band K. Let $\delta\colon K \to \underline{R}_i e$ be a map so that $h\delta = he$ for $h \in \underline{R}_i$, δ maps $K \backslash \underline{R}_i(R \backslash R_i)$ injectively into $R_i e \backslash R_i(R_1 \backslash R_i)$, and $k \; \mathcal{R} \; k\delta$ for all $k \in K$. Put $1\delta = 1$ where 1 is an identity for R. Let $R(K \backslash \underline{R}_i)$ be a set of formal products. For $m = |K \backslash \underline{R}_i(R \backslash R_i)|_r$ define $V(R,I,M) = R(K \cup \{1\})$ so that $(rh)(sk) = r(h\delta)sk$ for $r,s \in R$, $h,k \in K \cup \{1\}$. Since K embeds $\underline{R}_i$ and $eR = R$ we see that $V(R,i,m)$ is a band embedding $R \cup K$.

LEMMA 2.11. *Suppose* $\underline{R}_j$ *is a right zero semigroup and* $\underline{R}_j R_j = R_j$ *for each* j, $0 \leqslant j < i$. *Then* $V(R,i,m)$ *is an* F_R-*projective for* $m \leqslant m_0$ *if (i)* $\underline{R}_i$ *is a right zero semigroup with* $\underline{R}_i R_i = R_i$ *or*

(ii) $m_{i-1} = 1$, $R_i \underline{R}_i = R_i$ *and either* $\underline{R}_i R_i \neq R_i$ *or* $\underline{R}_i$ *is not a right zero semigroup.*

It can be shown that the process of building F_R-projectives from an F_R-projective with an identity, using lemmas 2.10 and 2.11, is the reverse of the reduction of theorem 2.7(i). Furthermore it has been noted that the one element semigroup is an F_R-projective. By an induction argument on the number of principal factors of an F_R-projective that utilises theorem 2.7 and lemmas 2.9, 2.10 and 2.11 we get

THEOREM 2.12. *Any F_R-projective is the result of a succession of extensions, beginning with the trivial semigroup, of the types described in lemmas 2.9, 2.10, 2.11 and their duals.*

By lemma 2.5 any rectangular band that is an F_R-projective is a finite right or left zero semigroup. Conversely, by lemma 2.10, any finite right or left zero semigroup can be seen to be an F_R-projective.

Suppose R is an F_R-projective that is a chain of two rectangular bands $\underline{R}_0$ and $\underline{R}_1$. By theorem 2.12 R is of one of the following forms, or their dual forms. By lemma 2.5 we can assume $\underline{R}_0 = \{e_1,\ldots,e_n\}$ is a right zero semigroup. Then we have one of the following cases.

Case (a): $\underline{R}_1 = \{f_{ij};\ 1 \leqslant i \leqslant r,\ 1 \leqslant j \leqslant n\}$ is a rectangular band with L-classes $\{f_{ij};\ 1 \leqslant i \leqslant r\}$, so that $e_j f_{ij} e_j = f_{ij}$.

Case (b): $\underline{R}_1 = \{f_{ij}, f_{n+h};\ 1 \leqslant i \leqslant r, 1 \leqslant j \leqslant n, 1 \leqslant h \leqslant m\}$ is a right zero semigroup with $r \geqslant m$, so that $e_j f_{ij} e_j = f_{ij}$ and $f_{n+h} e_j = f_{hj}$.

Case (c): $\underline{R}_1 = \{f_{ik};\ 1 \leqslant i \leqslant r, 1 \leqslant k \leqslant n+1\}$ is a rectangular band with L-classes $\{f_{ik};\ 1 \leqslant i \leqslant r\}$ and so that $e_j f_{ik} e_j = f_{ij},\ 1 \leqslant j \leqslant n$.

As promised before lemma 2.9, we will now check that these are in fact F_R-projectives. Use is made of the notion of sandwich sets. For $x,y \in E_T$, the set of idempotents of a regular semigroup T, the *sandwich set* is

$$S(x,y) = \{z \in E_T\colon zx = z = yz,\ xzy = xy\}.$$

By [Nambooripad 1974b] we have

LEMMA 2.13. *Let $\beta\colon T \to S$ be a homomorphism of regular semigroups, and $x,y \in E_T$. Then (i) $S(x,y) \neq \square$; (ii) $(S(x,y))\beta \subseteq S(x\beta,y\beta)$; and (iii) $x\,\mathcal{R}\,y$ implies $S(x,y) = \{x\}$.*

For R as in cases (a), (b) or (c), let $\alpha\colon R \to S$ be a morphism and $\beta\colon T \to S$ be an epimorphism. By [Hall 1972] we may choose $\{g_1,\ldots,g_n\}$ to be a right zero subsemigroup of T so that $g_j\beta = e_j\alpha,\ 1 \leqslant j \leqslant n$.

Suppose R is as in case (a). The left zero semigroup with adjoined identity $R' = \{e_1\} \cup \{f_{i1};\ 1 \leq i \leq r\}$ is clearly an F_R-projective. We can check that g_1Tg_1 and $(e_1\alpha)S(e_1\alpha)$ are regular semigroups and $(g_1Tg_1)\beta = (e_1\alpha)S(e_1\alpha)$. By the definition of F_R-projectives there is a morphism $\gamma'\colon R' \to T$ so that $\gamma'\beta = \alpha$ when restricted to R' and $e_1\gamma' = g_1$. Extend γ' to $\gamma\colon R \to T$ by defining $e_j\gamma = g_j$ and $f_{ij}\gamma = f_{i1}\gamma' g_j$ for all i and j. It is easily seen that γ is a morphism and $\gamma\beta = \alpha$.

Now suppose R is as in case (b). Choose a set of idempotents $\{q_{ij}, q_{n+h};\ 1 \leq i \leq r, 1 \leq j \leq n, 1 \leq h \leq m\}$ in T so that $q_{ij}\beta = f_{ij}\alpha$ and $q_{n+h}\beta = f_{n+h}\alpha$. In particular choose $q_{11} \leq g_1$; this choice can be made by considering β restricted to the regular subsemigroup g_1Tg_1 of T, as in the proof of theorem 1.3. Let $p_1 = q_{11}$ and define inductively $p_i \in S(q_{i1}, p_{i-1}\cdots p_1)$ and $p_{n+h} \in S(q_{n+h}, p_{n+h-1}\cdots p_1)$. Note that by lemma 2.13 $p_2\beta = f_{21}\alpha$. Also $p_1p_2 = p_2 = g_jp_2$, so $p_2\{p_1, g_j;\ 1 \leq j \leq n\}$ is a right zero subsemigroup of T. More generally $p_i\beta = f_{i1}\alpha$ and $p_{n+h}\beta = f_{n+h}\alpha$. Also since $p_i = p_{i-1}\cdots p_1p_i$, $1 \leq i \leq n + m$, then $g_jp_i = p_{i-1}\cdots p_kp_i = p_i$, $1 \leq k < i$, so $p_{n+m}\{g_i, p_{n+m}\cdots p_k;\ 1 \leq i \leq n, 1 \leq k \leq n + m\}$ is a right zero subsemigroup of T. Now define

$$d_{n+h} = p_{n+m}\cdots p_{n+h} \quad \text{for} \quad 1 \leq h \leq m \quad \text{and}$$

$$d_{ij} = \begin{cases} p_{n+m}\cdots p_{n+i}g_j & \text{if } 1 \leq i \leq m \\ p_{n+m}\cdots p_ig_j & \text{if } m < i \leq r. \end{cases}$$

It is easily seen that $d_{ij}\beta = f_{ij}\alpha$ and $d_{n+h}\beta = f_{n+h}\alpha$. Define $\gamma\colon R \to T$ by $e_j\gamma = g_j$, $f_{ij}\gamma = d_{ij}$ and $f_{n+h}\gamma = d_{n+h}$, for $1 \leq i \leq r$, $1 \leq j \leq n$ and $1 \leq h \leq m$. Then $\gamma\beta = \alpha$, and γ is a homomorphism.

Finally suppose R is as in case (c). Let $\{q_{ik};\ 1 \leq i \leq r, 1 \leq k \leq n + 1\}$ be a subset of idempotents of T so that $q_{ik}\beta = f_{ik}\alpha$. As in the last case assume $q_{11} \leq g_1$. Choose $p \in S(q_{1(n+1)}, q_{11})$. Then $\{p\} = \{g_i;\ 1 \leq i \leq n\}p$ and $p\beta = f_{1(n+1)}\alpha$. Choose $p_2 \in S(pg_1, q_{21})$, so $p_2\beta = f_{21}\alpha$ and $g_1p_2p\{g_j;\ 1 \leq j \leq n\}$ is a right zero subsemigroup of T. Continuing this selection, choose $p_i \in S(g_1p_2\cdots p_{i-1}, q_{i1})$. So $p_i\beta = f_{i1}\alpha$ and $p_ig_1p_2\cdots p_{i-1} = p_ig_1p_h\cdots p_{i-1} = p_ipg_1$ for $2 \leq h \leq i$. From these relations we can see that the following elements form a rectangular band: for $1 \leq j \leq n$, $1 \leq i \leq r$, define

$$d_{1j} = pg_1p_2\dots p_rpg_j; \quad d_{1(n+1)} = pg_1p_2\dots p_rp$$

$$d_{ij} = g_1p_i\dots p_rpg_j; \quad d_{i(n+1)} = g_1p_i\dots p_rp.$$

We have $d_{ik}\beta = f_{ik}\alpha$ for $1 \leqslant k \leqslant n + 1$. Define the homomorphism $\gamma: R \to T$ by $f_{ik}\gamma = d_{ik}$ and $e_i\gamma = g_i$. Then $\gamma\beta = \alpha$.

It should be noted that the proofs of lemmas 2.9, 2.10 and 2.11 can be used in the category of regular semigroups. Thus theorem 2.12 also describes a process by which all finite weak projectives can be determined in the category of regular semigroups. Similarly the F_R-projectives are weak band projectives; so we have enlarged the class of known weak band projectives (the characterizations of [Nordahl and Scheiblich, B] are not exclusive).

3. Inverse semigroups

In the category I of inverse semigroups a different approach can be used in characterizing the projectives, since epimorphisms in I are surjective morphisms and free inverse semigroups exist. The I-projectives are then the retracts of free inverse semigroups.

The I-projectives can be easily seen, therefore, to be both fundamental and E-unitary. In [Trotter, C] the description of free inverse semigroups by Scheiblich [1973b] and of E-unitary inverse semigroups by McAlister [1974a] are utilized to give

THEOREM 3.1. *Let S be the inverse subsemigroup of the free inverse product $I(Y) * E(Z)$ of the free inverse semigroup $I(Y)$ and the free semilattice $E(Z)$ which is generated by $\{w(x): x \in Y \cup Z\}$ where for each $x \in Y \cup Z$,*

(a) $w(x) = e(x)xf(x)$ for some idempotents $e(x), f(x) \in (I(Y)* E(Z))^1$ and

(b) $(x_1\dots x_n)(x_1\dots x_n)^{-1} \geqslant w(x)(w(x))^{-1}$, $x_1,\dots,x_n \in Y \cup Y^{-1} \cup Z$, implies $x_1\dots x_{n-1}w(x_n)(x_1\dots x_{n-1}w(x_n))^{-1} \geqslant w(x)(w(x))^{-1}$.

Then S is an I-projective; $\{w(x): x \in Y \cup Z\}$ is a basis for S if and only if *(c)* $J_{w(x)} \neq J_{w(z)}$ *for all* $x \neq z \in Z$.

Conversely any I-projective is isomorphic to such a semigroup.

For S as in the theorem, the free group $G(Y)$ on Y is the maximal group homomorphic image of S.

References

1954 B.H. Neumann, An essay on free products of groups with amalgamation. *Philos. Trans. Roy. Soc. London* Ser.A 246 (1954), 503-554.

1966 S. Lang, *Rapport sur la cohomologie des groupes*. Benjamin, New York 1966.

1967 S. Mac Lane and G. Birkhoff, *Algebra*. Macmillan, New York 1967.

1971 A. Horn and N. Kimura, The category of semilattices. *Algebra Universalis* 1 (1971), 26-38.

1972 T.E. Hall, Congruences and Greens relations on regular semigroups. *Glasgow Math. J.* 13 (1972), 167-175.

1973a M. Petrich, *Introduction to semigroups*. Merrill, Columbus 1973.

1973b H.E. Scheiblich, Free inverse semigroups. *Proc. Amer. Math. Soc.* 38 (1973), 1-7.

1974a D.B. McAlister, Groups, semilattices and inverse semigroups II. *Trans. Amer. Math. Soc.* 196 (1974), 351-370.

1974b K.S.S. Nambooripad, Structure of regular semigroups I. *Semigroup Forum* 9 (1974), 354-363.

1978 T.E. Hall, Amalgamation and inverse and regular semigroups. *Trans. Amer. Math. Soc.* 246 (1978), 395-406.

TO BE PUBLISHED

[A] P.R. Jones, Universal aspects of completely simple semigroups. These proceedings.

[B] T.E. Nordahl and H.E. Scheiblich, Projective bands. *Algebra Universalis* (to appear).

[C] P.G. Trotter, Projectives in inverse semigroups. Submitted for publication.

GENERALIZED INVERSE SEMIGROUPS AND AMALGAMATION

T. E. Hall

Department of Mathematics, Monash University, Clayton,
Victoria, Australia, 3168.

Abstract

Given that free products with amalgamation of inverse semigroups exist without "collapse", or equivalently that any amalgam of inverse semigroups is strongly embeddable in an inverse semigroup, it is natural to ask likewise if free products with amalgamation of generalized inverse semigroups exist without collapse. We note the result of Imaoka [1976b], that free products exist for the class of generalized inverse semigroups.

As yet we are unable to answer this question. The main result of this paper is that any amalgam of generalized inverse semigroups is strongly embeddable in a semigroup. Of course this gives some hope that our question above will have an affirmative answer. Our proof is via three representation extension properties. We show that any generalized inverse semigroup has the representation extension property in any containing generalized inverse semigroup; and that any right generalized inverse semigroup has the free (and hence the strong) representation extension property in any containing right generalized inverse semigroup. Our results have been obtained independently by Teruo Imaoka (private communication).

1. Preliminaries

An *amalgam of semigroups* is a family $(S_i, i \in I; U)$ of semigroups such that U is a subsemigroup of each S_i and such that $S_i \cap S_j = U$ for all distinct $i, j \in I$. An amalgam $(S_i, i \in I; U)$ is *embeddable in a*

ISBN 0-12-319450-4

semigroup W if there exist monomorphisms $\phi_i: S_i \to W$ $(i \in I)$ agreeing precisely on U, i.e. $\phi_i|U = \phi_j|U$ and $(S_i\phi_i) \cap (S_j\phi_j) = U_i\phi_i$ for all distinct $i,j \in I$.

A *generalized inverse semigroup* is a regular semigroup in which the idempotents satisfy the permutation identity $efgh = egfh$ [Yamada, 1967], while a *right* [*left*] *generalized inverse semigroup* is a regular semigroup in which $efg = feg$ [$efg = egf$] for all idempotents e,f,g.

The semigroup of all (total) transformations of a set X under composition is denoted by $T(X)$. By a *representation* of a semigroup S is meant any morphism $\rho: S \to T(X)$, for any set X.

Representation extension has been found to be closely connected with the embeddability of semigroup amalgams. A semigroup U is said to have the *representation extension property* in a containing semigroup S if for every representation $\rho: U \to T(X)$ of U, there exist a set Y and a representation $\alpha: S \to T(X \cup Y)$ such that $\alpha_u|X = \rho_u$ for all $u \in U$. By theorem 7 of the author [1978] if, for every semigroup T containing U as a subsemigroup, the amalgam $(S,T;U)$ is embeddable then U has the representation extension property in S; thus in determining whether or not an amalgam $(S_i, i \in I; U)$ is embeddable, it has often helped to determine first whether or not U has the representation extension property in S_i, for each $i \in I$.

A semigroup U is said to have the *strong representation extension property* in a containing semigroup S if for any representations $\rho: U \to T(X)$, $\beta: S \to T(X')$ such that $X' \subseteq X$ and $\rho_u|X' = \beta_u$ for all $u \in U$, there exist a set Y and a representation $\alpha: S \to T(X \cup Y)$ such that $\alpha_u|X = \rho_u$ and $\alpha_s|X' = \beta_s$ for all $u \in U$, $s \in S$.

RESULT 1 ([the author, 1978], theorem 4). Let $(S_i, i \in I; U)$ be any amalgam of semigroups. If U has the strong representation extension property in each S_i then the amalgam is embeddable.

A previous example for result 1 is given by theorem 6 of [the author, 1978], namely that any inverse semigroup U has the strong representation extension property in any containing semigroup S. We obtain a further example in this paper by showing that any right generalized inverse semigroup U has the strong representation extension property in any containing right generalized inverse semigroup S. In fact we simultaneously generalize both of these mentioned cases by showing that a regular semigroup U has the strong representation extension property in a containing semigroup S if $efs = fes$ for all $e,f \in E(U)$ (the set of idempotents of U) and for all $s \in S$ (note that if S is a right generalized inverse semigroup then $efs = fes$ for all $e,f \in E(S)$, $s \in S$).

2. The free extension of a representation

In this section we give some further preliminary concepts and results, taken from [the author, 1978].

Take any subsemigroup U of a semigroup S and any representation $\rho: U \to T(X)$. First of all, we shall see a manipulative and so far quite successful way of proving that the representation $\rho: U \to T(X)$ extends to one of S. Produced via the concept of the free extension of ρ to S, the manipulation consists of showing that a set of zigzag equations imply the equality of two elements in X.

Let $S^{(1)}$ denote the semigroup S with an identity element 1 adjoined whether or not S has an identity, and likewise for $U^{(1)}$ ($\leqslant S^{(1)}$). On $X \times S^{(1)}$ define a relation δ_1 as follows: for all $(x,s),(y,t) \in X \times S^{(1)}$, $(x,s)\delta_1(y,t)$ if and only if $s = us_1$, $t = vs_1$, $x\rho_u = y\rho_v$ for some $u,v \in U^{(1)}$, $s_1 \in S^{(1)}$ (we define $x\rho_1 = x$ for all $x \in X$). The transitive closure of δ_1 we denote by δ, i.e. $\delta = \cup_{n=1}^{\infty} \delta_1^n$, an equivalence relation on $X \times S^{(1)}$. We define a representation $\alpha: S \to T((X \times S^{(1)})/\delta)$ as follows: for all $(x,t) \in X \times S^{(1)}$, $s \in S$, $(x,t)\delta\alpha_s = (x,ts)\delta$ (each α_s is well-defined). To show that ρ extends to a representation of S is equivalent to showing that the map $f: X \to (X \times S^{(1)})/\delta$, $x \mapsto (x,1)\delta$ is one-to-one. The ordered pair (α,f) is called the *free extension of* ρ *to* S (though f is not always one-to-one).

Henceforth we denote $x\rho_u$ by xu, for all $x \in X$, $u \in U^{(1)}$. For any $(x,s),(y,t) \in X \times S^{(1)}$, the condition $(x,s)\delta(y,t)$ is equivalent to a set of zigzag equations obtained as follows. There exists a positive integer n such that $(x,s)\delta_1^n(y,t)$ and so there exist elements $x_1,x_2,\ldots,x_{n+1} \in X$, $u_1,v_1,u_2,v_2,\ldots,u_n,v_n \in U^{(1)}$ and $s_1,s_2,\ldots,s_n \in S^{(1)}$ such that

$$(x,s) = (x_1,u_1s_1),(x_{n+1},v_ns_n) = (y,t),$$

$$x_1u_1 = x_2v_1, x_2u_2 = x_3v_2,\ldots,x_nu_n = x_{n+1}v_n, \tag{1}$$

and

$$v_1s_1 = u_2s_2, v_2s_2 = u_3s_3, \ldots, v_{n-1}s_{n-1} = u_ns_n. \tag{2}$$

In the case where $s = t = 1$ we have $u_1 = s_1 = v_n = s_n = 1$. Hence showing that f is one-to-one, and thus showing that U has the representation extension property in S, is equivalent to showing that the equations (1) and (2), together with $u_1 = s_1 = v_n = s_n = 1$, imply that $x_1 = x_{n+1}$. Lemmas 13 and 14 of [the author, 1978] further show that we may assume without loss of generality that none of $u_2, u_3, \ldots, u_n$, $v_1, v_2, \ldots, v_{n-1}, s_2, s_3, \ldots, s_{n-1}$ equal 1 (when proving f one-to-one).

Now we shall see how proving the strong representation extension property can be reduced to showing two conclusions from the zigzag equations (1) and (2), one being that given above.

Given $\rho\colon U^{(1)} \to T(X)$ above, take any representation $\rho'\colon U^{(1)} \to T(X')$ such that $X' \subseteq X$ and $\rho_u|X = \rho'_u$ for all $u \in U^{(1)}$. We define δ'_1 on $X' \times S^{(1)}$ to be the restriction of δ_1 to $X' \times S^{(1)}$ and we define δ' to be the transitive closure of δ'_1. The free extension of ρ' to S is denoted by α'; then $(x,t)\delta'\alpha'_s = (x,ts)\delta'$ for all $(x,t) \in X' \times S^{(1)}$, $s \in S$. The *free representation extension property of U in S* is equivalent to the condition that for all ρ,ρ' as above δ' equals the restriction of δ to $X' \times S^{(1)}$, but is defined as being equivalent to the condition that for all ρ,ρ' as above, α' is equivalent to a subrepresentation of α under the map $(X' \times S^{(1)})/\delta' \to (X \times S^{(1)})/\delta$, $(x,s)\delta' \mapsto (x,s)\delta$ (see [the author, 1978] for details).

Showing the free representation extension property of U in S is thus equivalent to showing that equations (1) and (2), together with $x_1, x_{n+1} \in X'$ imply that $(x_1, u_1s_1)\delta'(x_{n+1}, v_ns_n)$. The following result shows that proving the strong representation extension property of U in S is equivalent to proving both the conclusions given above.

RESULT 2 (theorem 3 of [the author, 1978]). A subsemigroup U of a semigroup S has the strong representation extension property in S if and only if U has both the representation extension property and the free representation extension property in S.

We note that remark 2 of [the author, 1978] places the free extension of ρ to S as an example of a concept in universal algebra, together with, for example, the free group on a semigroup.

3. Embedding amalgams

Let now U be any regular subsemigroup of a semigroup S such that for all idempotents $e,f \in U$, for any element $s \in S$, we have $efs = fes$. For example, U could be an inverse subsemigroup of any semigroup S, or secondly, U and S could be right generalized inverse semigroups. We show that U has the strong representation extension property in S.

Take any representation $\rho\colon U \to T(X)$ and define δ, α, f as above. Take any $(x,s),(y,t) \in X \times S^{(1)}$ such that $(x,s)\delta(y,t)$. As before, there exist a positive integer n and elements $x_1, x_2, \ldots, x_{n+1} \in X$, $u_1, v_1, u_2, v_2, \ldots, u_n, v_n \in U^{(1)}$ and $s_1, s_2, \ldots, s_n \in S^{(1)}$ such that $(x,s) = (x_1, u_1 s_1), (x_{n+1}, v_n s_n) = (y,t)$ and equations (1) and (2) hold.

For $i = 1,2,\ldots,n$ let u_i', v_i' denote inverses in $U^{(1)}$ of u_i, v_i respectively, and put $w_i = u_1 v_1' u_2 v_2' \ldots u_i v_i'$ and $w_i' = v_i u_i' \ldots v_2 u_2' v_1 u_1'$, an inverse of w_i in $U^{(1)}$.

LEMMA 1. *(i) If $s_1, s_2, \ldots, s_n$ are each not equal to 1, then* $w_n' w_n v_n s_n = w_n' u_1 s_1$.

(ii) If $v_n = s_n = u_1 = s_1 = 1$ and if $s_2, s_3, \ldots, s_{n-1}$ are each not equal to 1, then $w_n' w_n = w_n'$.

(iii) $x_{n+1} w_n' = x_1 w_n w_n'$.

Proof. The proofs of the three parts are all similar.

$$\begin{aligned}
\text{(i)}\quad w_n' w_n v_n s_n &= v_n (u_n' w_{n-1}' w_{n-1} u_n)(v_n' v_n) s_n \\
&= v_n (v_n' v_n)(u_n' w_{n-1}' w_{n-1} u_n) s_n \\
&= w_n' w_{n-1} v_{n-1} s_{n-1} = \cdots = w_n' w_1 v_1 s_1 \\
&= v_n u_n' \ldots v_2 u_2' v_1 (u_1' u_1)(v_1' v_1) s_1 \\
&= v_n u_n' \ldots v_2 u_2' v_1 (v_1' v_1)(u_1' u_1) s_1 \\
&= w_n' u_1 s_1 .
\end{aligned}$$

(ii) The equations in the proof of part (i) are also true under the assumptions of part (ii).

(iii) $x_{n+1}w'_n = x_{n+1}v_nv'_nw'_n = x_nu_nv'_nw'_n = x_n(u_nv'_nv_nu'_n)(v_{n-1}v'_{n-1})w'_{n-1}$

$= x_n(v_{n-1}v'_{n-1})(u_nv'_nv_nu'_n)w'_{n-1}$ (since $v_{n-1}, w'_{n-1} \in S$ or $v_{n-1}v'_{n-1} = 1$)

$= x_{n-1}u_{n-1}v'_{n-1}u_nv'_nw'_n = \ldots = x_1u_1v'_1\ldots u_nv'_nw'_n$

$= x_1w_nw'_n$.

THEOREM 2. *Let U be a regular subsemigroup of a semigroup S such that $efs = fes$ for all $e,f \in E(U)$, $s \in S$. Then U has the representation extension property in S.*

Proof. Take any representation $\rho\colon U \to T(X)$ of U and define δ_1 and δ on $X \times S^{(1)}$ as above. Take any $x,y \in X$ such that $(x,1)\delta = (y,1)\delta$. We need only show that $x = y$.

As before there exists a positive integer n such that $(x,1)\delta_1^n(y,1)$ and there exist elements $x_1,x_2,\ldots,x_{n+1} \in X$, $u_1,v_1,u_2,v_2,\ldots,u_n,v_n \in U^{(1)}$ and $s_1,s_2,\ldots,s_n \in S^{(1)}$ such that equations (1) and (2) hold.

We proceed by induction on n. If $(x,1)\delta_1(y,1)$ then $x = y$ since $u_1s_1 = v_1s_1 = 1$ implies $u_1 = v_1 = 1$. We assume then that for all m, $1 \leqslant m < n$, and for all $x',y' \in X$, if $(x',1)\delta_1^m(y',1)$ then $x' = y'$.

Case I; $s_i = 1$ for some $i \in \{2,3,\ldots,n-1\}$.

Now $(x,1)\delta_1^{i-1}(x_i,u_is_i)\delta_1(x_iu_i,1)$, whence $x = x_iu_i$. Likewise $(x_iu_i,1) = (x_iu_i,s_i)\delta_1(x_{i+1},v_is_i)\delta_1^{n-i}(y,1)$, whence $x_iu_i = y$, giving $x = y$ as required. (This is essentially the proof of lemma 13 of [the author, 1978] mentioned above.)

Case II; none of $s_2,s_3,\ldots,s_{n-1}$ equal 1.

The proof now goes in the three stages of the proof of theorem 1 of the author [1978]; as in the proof of lemma 1 the elements $s_2,s_3,\ldots,s_{n-1} \in S$ allow "permuting of idempotents from U in certain products from S". We have

$x = x_1 = x_2v_1 = x_2(w'_1w_1v_1s_1) = x_2(w'_1w_1)(u_2s_2) = x_2u_2u'_2(w'_1w_1)(u_2s_2) =$
$x_3v_2u'_2(w'_1w_1)(u_2s_2) = x_3(v_2(v'_2v_2)(u'_2w'_1w_1u_2)s_2) = x_3(v_2(u'_2w'_1w_1u_2)$
$(v'_2v_2)s_2) = x_3(w'_2w_2v_2s_2) = \ldots = x_{n+1}w'_nw_nv_ns_n = x_{n+1}w'_nw_n$

(end of stage 1)

$= x_{n+1}w'_n$ (from lemma 1(ii)) (end of stage 2)

$= x_1w_nw'_n$ (from lemma 1(iii)) (end of stage 3)

$= y,$ (from symmetry and $x = x_{n+1}w'_nw_n$)

completing the proof.

THEOREM 3. *Let U be a regular subsemigroup of a semigroup S such that $efs = fes$ for all $e,f \in E(U)$, $s \in S$. Then U has the free representation extension property in S.*

Proof. Take any representations $\rho: U^{(1)} \to T(X)$, $\rho': U^{(1)} \to T(X')$ such that $X' \subseteq X$, $\rho_u|X' = \rho'_u$ for all $u \in U^{(1)}$, and $\rho_1 = \iota_X$. Define δ_1, δ on $X \times S^{(1)}$ and δ'_1, δ' on $X' \times S^{(1)}$ as above and take any elements $x,y \in X'$, $s,t \in S^{(1)}$ such that $(x,s)\delta(y,t)$. We need only show that $(x,s)\delta'(y,t)$.

As before there exist a positive integer n and elements $x_1,x_2,\ldots,x_{n+1} \in X$, $u_1,v_1,u_2,v_2,\ldots,u_n,v_n \in U^{(1)}$, $s_1,s_2,\ldots,s_n \in S^{(1)}$ such that $(x,s) = (x_1,u_1s_1),(x_{n+1},v_ns_n) = (y,t)$ and such that equations (1) and (2) hold.

Case 1; some of $s_1,s_2,\ldots,s_n$ equal 1.

Let i be the smallest of $1,2,\ldots,n$ such that $s_i = 1$. We show first that $(x_1,u_1s_1)\delta'(x_iw'_{i-1}w_{i-1}u_i,1)$. We have of course that $s_1,s_2,\ldots,s_{i-1} \in S$ so

$$(x_1,u_1s_1)\delta'_1(x_1u_1,s_1) = (x_2v_1u'_1u_1,s_1)\delta'_1(x_2v_1,(v'_1v_1)(u'_1u_1)s_1)$$

$$= (x_2v_1,(u'_1u_1)(v'_1v_1)s_1) = (x_2v_1,u'_1u_1v'_1u_2s_2)\delta'_1(x_2v_1u'_1u_1v'_1u_2,s_2)$$

$$= (x_2w'_1w_1u_2,s_2) = (x_2(w'_1w_1)(u_2u'_2)u_2,s_2)$$

$$= (x_2(u_2u'_2)(w'_1w_1)u_2,s_2) \qquad \text{(since } u_2 \in U \text{ or } u_2u'_2 = 1\text{)}$$

$$= (x_3v_2u'_2w'_1w_1u_2,s_2)\delta'_1(x_3v_2u'_2w'_1w_1u_2,(v'_2v_2)(u'_2w'_1w_1u_2)s_2)$$

$$= (x_3v_2u'_2w'_1w_1u_2,(u'_2w_1w_1u_2)(v'_2v_2)s_2)$$

$$= (x_3v_2u'_2w'_1w_1u_2,u'_2w_1w_1u_2v'_2u_3s_3)\delta'_1(x_3v_2u'_2w'_1w_1u_2v'_2u_3,s_3)$$

$$= (x_3w'_2w_2u_3,s_3) = \ldots = (x_iw'_{i-1}w_{i-1}u_i,s_i) = (x_iw'_{i-1}w_{i-1}u_i,1).$$

Now let k be the largest of $1,2,\ldots,n$ such that $s_k = 1$. By symmetry, we have that $(y,t)\delta'(x_k u,1)$ for some $u \in U^{(1)}$. Now $(x_k u,1)\delta'(y,t)\delta(x,s)\delta'(x_i w_{i-1}' w_{i-1} u_i,1)$ so $(x_k u,1)\delta(x_i w_{i-1}' w_{i-1} u_i,1)$ (since $\delta' \subseteq \delta$) whence $x_k u = x_i w_{i-1}' w_{i-1} u_i$ from the proof of theorem 2. Now we have that

$$(x,s)\delta'(x_i w_{i-1}' w_{i-1} u_i,1) = (x_k u,1)\delta'(y,t)$$

giving $(x,s)\delta'(y,t)$ as required.

Case 2; $s_1,s_2,\ldots,s_n \in S$.

The proof now goes in the three stages of the proof of theorem 2 of the author [1978]; again the elements $s_1,s_2,\ldots,s_n \in S$ allow "permuting of idempotents from U in certain products from S". We have

$(x_1,u_1s_1)\delta'(x_n w_{n-1}' w_{n-1} u_n, s_n)$ (as in the proof for Case 1)

$= (x_n u_n u_n' w_{n-1}' w_{n-1} u_n, s_n)$ ($u_n \in U$ or $u_n u_n' = 1$)

$= (x_{n+1} v_n u_n' w_{n-1}' w_{n-1} u_n, s_n)$

$\delta_1'(x_{n+1}, v_n(v_n' v_n)(u_n' w_{n-1}' w_{n-1} u_n)s_n)$ ($x_{n+1} \in X'$)

$= (x_{n+1}, v_n(u_n' w_{n-1}' w_{n-1} u_n)(v_n' v_n)s_n)$

$= (x_{n+1}, w_n' w_n v_n s_n)$ (end of stage 1)

$= (x_{n+1}, w_n' u_1 s_1)$ (by lemma 1(i)) (end of stage 2)

$\delta_1'(x_{n+1} w_n', u_1 s_1)$

$= (x_1 w_n w_n', u_1 s_1)$ (by lemma 1(iii))

$\delta_1'(x_1, w_n w_n' u_1 s_1)\delta'(y,t)$ from symmetry and since

$(x,s)\delta'(x_{n+1}, w_n' w_n v_n s_n)$.

This completes the proof of theorem 3.

THEOREM 4. *If U is a regular subsemigroup of a semigroup S such that $efs = fes$ for all $e,f \in E(U)$, $s \in S$ then U has the strong representation extension property in S.*

Proof. This follows from result 2 and theorems 2 and 3.

COROLLARY 5. *An amalgam $(S_i, i \in I; U)$ is embeddable if for each $i \in I$, $efs = fes$ for all $e,f \in E(U)$, $s \in S_i$.*

Proof. This follows from result 1 and theorem 4.

COROLLARY 6 [J.M. Howie, 1975]. *An amalgam $(S_i, i \in I; U)$ is embeddable if U is an inverse semigroup.*

COROLLARY 7. *An amalgam $(S_i, i \in I; U)$ of right generalized inverse semigroups is embeddable.*

A subsemigroup U of a semigroup S is said to have the [*right*] *congruence extension property* in S if for every [right] congruence ρ on U there is a [right] congruence δ on S such that $\delta \cap (U \times U) = \rho$.

COROLLARY 8. *If U is a regular subsemigroup of a semigroup S such that $efs = fes$ for all $e,f \in E(U)$, $s \in S$, then U has the right congruence extension property in S.*

Proof. This follows from theorem 9 of [the author, 1978] and theorem 2.

THEOREM 9. *Any amalgam $(S_i, i \in I; U)$ of generalized inverse semigroups is embeddable.*

Proof. For simplicity, let us deal with the case $|I| = 2$ first.

Thus we let $(S,T;U)$ be an amalgam of generalized inverse semigroups. From the excellent structure theorem for generalized inverse semigroups due to Yamada [1967] it follows that there exist left generalized inverse semigroups S_1, T_1, U_1 and right generalized inverse semigroups S_2, T_2, U_2, and monomorphisms $\phi: S \to S_1 \times S_2$, $\psi: T \to T_1 \times T_2$ such that $U_1 \leqslant S_1$, $U_1 \leqslant T_1$, $S_1 \cap T_1 = U_1$, $U_2 \leqslant S_2$, $U_2 \leqslant T_2$, $S_2 \cap T_2 = U_2$, $\phi|U = \psi|U$ and $U\phi \subseteq U_1 \times U_2$.

Now the amalgam $(S_2, T_2; U_2)$ is embeddable in a semigroup W_2 say, by corollary 7. Dually, the amalgam $(S_1, T_1; U_1)$ is embeddable in a semigroup W_1, say. Clearly now, the amalgam $(S,T;U)$ is embeddable in the semigroup $W_1 \times W_2$.

The arbitrary case is dealt with quite similarly; one merely represents U and each S_i as a subdirect product of a left and a right generalized inverse semigroup.

4. Miscellaneous results

We consider in this section the questions of whether or not U has the [strong] representation extension property in S when S and U are both generalized inverse semigroups.

The answer is in the negative for the strong representation extension property. We let $U = \{a,b\}$ be the two-element left zero semigroup and put $S = U^0$. The representation $\rho\colon U \to T(X)$ given in the proof of theorem 25(v) of [the author, 1978] shows that U does not have the strong representation extension property in S. Note that S is in fact left generalized inverse.

The answer is in the affirmative for the representation extension property. The author obtained the proof of the following theorem by generalizing an ingenious proof of P.R. Jones that a normal band U has the right congruence extension property in a containing normal band S first to a proof of the same result for generalized inverse semigroups and then to the following proof.

THEOREM 10. *Let U and S be any generalized inverse semigroups with U a subsemigroup of S. Then U has the representation extension property in S.*

Proof. Take any semigroup T containing U as a subsemigroup and any zigzag of type II(a) (see [the author, 1978]) over U from an element $t \in T$ to an element $t' \in T$, say

$$t = t_1u_1,\ t_nv_n = t',$$

$$u_1 = v_1s_1,\ u_2s_1 = v_2s_2,\ u_3s_2 = v_3s_3,\ \ldots,\ u_ns_{n-1} = v_n,$$

$$t_1v_1 = t_2u_2,\ t_2v_2 = t_3u_3,\ \ldots,\ t_{n-1}v_{n-1} = t_nu_n,$$

where $s_1,\ldots,s_{n-1} \in S$, $t_1,\ldots,t_n \in T$, $u_1,v_1,\ldots,u_n,v_n \in U$. We need only show that $t = t'$, from which theorem 10 follows, by theorem 16 of [the author, 1978].

Let us further put

$$x_1 = u_1 = v_1s_1,\ldots,x_i = u_is_{i-1} = v_is_i,\ldots,x_n = u_ns_{n-1} = v_n,$$

and for $i = 1,2,\ldots,n-1$, we take any $s_i' \in V(s_i)$ and $v_i' \in V(v_i) \cap U$, and we put $x_i' = s_i'v_i'$, an inverse of x_i in S. We prove two lemmas before completing the proof of theorem 10.

LEMMA 11. *For* $i = 1, 2, \ldots, n$,

$$x_i(x'_{i-1}x_{i-1})\ldots(x'_2x_2)(x'_1x_1) = u_i(v'_{i-1}u_{i-1})\ldots(v'_2u_2)(v'_1u_1) \in U$$

and

$$t = t_i[x_i(x'_{i-1}x_{i-1})\ldots(x'_2x_2)(x'_1x_1)].$$

Proof. For $i = 1$, we have $x_1 = u_1 \in U$ and $t = t_1u_1 = t_1x_1$ as required. Suppose then that for some $k \in \{1,2,\ldots,n-1\}$ we have $x_k(x'_{k-1}x_{k-1})\ldots(x'_1x_1) = u_k(v'_{k-1}u_{k-1})\ldots(v'_2u_2)(v'_1u_1) \in U$ and $t = t_k[x_k(x'_{k-1}x_{k-1})\ldots(x'_1x_1)]$. Now

$$\begin{aligned} x_{k+1}x'_kx_k &= (u_{k+1}s_k)(s'_kv'_k)(v_ks_k) \\ &= u_{k+1}(s_ks'_k)(v'_kv_k)s_k \\ &= u_{k+1}(v'_kv_k)(s_ks'_k)s_k = u_{k+1}v'_kx_k \end{aligned}$$

whence

$$\begin{aligned} &x_{k+1}(x'_kx_k)\ldots(x'_2x_2)(x'_1x_1) \\ &= u_{k+1}v'_k[x_k(x'_{k-1}x_{k-1})\ldots(x'_2x_2)(x'_1x_1)] \\ &= u_{k+1}(v'_ku_k)(v'_{k-1}u_{k-1})\ldots(v'_2u_2)(v'_1u_1) \in U. \end{aligned}$$

Also

$$\begin{aligned} t &= t_k[x_k(x'_{k-1}x_{k-1})\ldots(x'_2x_2)(x'_1x_1)] \\ &= t_k[v_kv'_kx_k(x'_{k-1}x_{k-1})\ldots(x'_2x_2)(x'_1x_1)] \\ &= t_kv_kv'_ku_k(v'_{k-1}u_{k-1})\ldots(v'_2u_2)(v'_1u_1) \\ &= t_{k+1}u_{k+1}(v'_ku_k)\ldots(v'_2u_2)(v'_1u_1) \\ &= t_{k+1}[x_{k+1}(x'_kx_k)\ldots(x'_2x_2)(x'_1x_1)]. \end{aligned}$$

The lemma now follows by finite induction.

Denote the normal band of all idempotents of S by E. Choose any $x''_i \in V(x_i)$, $i = 1,2,\ldots,n$, and denote the J-class of E containing $(x''_nx_n)\ldots(x''_2x_2)(x''_1x_1)$ by J' (of course J' does not depend upon the choices of $x''_1, x''_2, \ldots, x''_n$). Put $B = \cup\{J \in E/J\colon J \geqslant J'\}$, a subband of E, and define a mapping $\theta\colon B \to J'$ by, for all $b \in B$, $b\theta = bjb$, where j is any element in J' (since B is normal, bjb does not depend on the choice of $j \in J'$, and further θ is a morphism).

LEMMA 12. *For any* $x_i'' \in V(x_i)$, $i = 1,2,\ldots,n$, *we have*

(i) $(x_1''x_1)\theta, (x_2''x_2)\theta, \ldots, (x_n''x_n)\theta$ *are* L*-related,*

(ii) $t = t_n(x_n[(x_n''x_n)\theta])$.

Proof. (i) For $i = 1,2,\ldots,n-1$, $x_i = v_is_i$ and $x_{i+1} = u_{i+1}s_i$ so $L_{x_i} \leqslant L_{s_i}$ and $L_{x_{i+1}} \leqslant L_{s_i}$ in S, i.e. $L_{x_i''x_i} \leqslant L_{s_i's_i}$ and $L_{x_{i+1}''x_{i+1}} \leqslant L_{s_i's_i}$ in S and hence in B. Since θ is a morphism and J' is a rectangular band, we have $(x_i''x_i)\theta$, $(x_{i+1}''x_{i+1})\theta$, and $(s_i's_i)\theta$ are all L-related in J', giving part (i).

(ii) $$t = t_n[x_n(x_n''x_n)(x_{n-1}'x_{n-1})\ldots(x_2'x_2)(x_1'x_1)] \quad \text{(lemma 11)}$$

$$= t_n[x_n[(x_n''x_n)(x_{n-1}'x_{n-1})\ldots(x_2'x_2)(x_1'x_1)]\theta]$$

$$= t_n[x_n[(x_n''x_n)\theta][(x_{n-1}'x_{n-1})\theta]\ldots[(x_2'x_2)\theta][(x_1'x_1)\theta]]$$

$$= t_n(x_n[(x_n''x_n)\theta]) \quad \text{(part (i))}.$$

We can now show that $t = t'$. First, we restrict the choice of x_1'', x_n'' in lemma 12 so that $x_1'' \in V(x_1) \cap U$, $x_n'' \in V(x_n) \cap U$. Now

$(x_1''x_1)\theta = (x_1''x_1)[x_n''(x_n(x_{n-1}'x_{n-1})\ldots(x_2'x_2)(x_1'x_1)](x_1''x_1) \in U$ and by symmetry $(x_n''x_n)\theta \in U$ also.

Lemma 12(ii) can be stated as

$$t = t'[(x_n''x_n)\theta]$$

and from symmetry we have

$$t' = t[(x_1''x_1)\theta]$$

$$= t'[(x_n''x_n)\theta][(x_1''x_1)\theta]$$

$$= t'[(x_n''x_n)\theta] \quad \text{(lemma 12(i))}$$

$$= t,$$

completing the proof of theorem 10.

A class of semigroups is said to have the [*right*] *congruence extension property* if for all semigroups U and S in the class, with U a subsemigroup of S, U has the [right] congruence extension property in S.

COROLLARY 13. *The class of generalized inverse semigroups has the right congruence extension property.*

Proof. This follows from theorem 9 of [the author, 1978] and theorem 10.

COROLLARY 14 (P.R. Jones, verbal communication, 1979). *The class of normal bands has the right congruence extension property.*

We note the result of Imaoka [1976a] that the class of normal bands has the congruence extension property. Of course the class of generalized inverse semigroups does not have the congruence extension property (since groups do not have the property).

The problem of whether an amalgam of generalized inverse semigroups is embeddable in a generalized inverse semigroup remains open. The reader may refer to the article of the author [A] for background and applications of amalgamation for regular and inverse semigroups.

References

1967 M. Yamada, Regular semigroups whose idempotents satisfy permutation identities. *Pacific J. Math.* 21 (1967), 371-392.

1975 J.M. Howie, Semigroup amalgams whose cores are inverse semigroups. *Quart. J. Math.* Oxford (2), 26 (1975), 23-45.

1976a T. Imaoka, Free products with amalgamation of bands. *Mem. Fac. Lit. and Sci.*, Shimane Univ. Nat. Sci. 10 (1976), 7-17.

1976b T. Imaoka, Free products of orthodox semigroups. *Mem. Fac. Lit. and Sci.*, Shimane Univ. Nat. Sci. 10 (1976), 1-6.

1978 T.E. Hall, Representation extension and amalgamation for semigroups. *Quart. J. Math.* Oxford (2), 29 (1978), 309-334.

TO BE PUBLISHED

[A] T.E. Hall, Inverse and regular semigroups and amalgamation: a brief survey. *Proceedings of the Conference on Regular Semigroups*, at De Kalb, in April 1979.

ON COMPLETELY REGULAR SEMIGROUP VARIETIES AND THE AMALGAMATION PROPERTY

G. T. Clarke

Department of Mathematics, Monash University, Clayton, Victoria, Australia, 3168.

Abstract

We consider the following question. Which varieties of completely regular semigroups have the weak or strong amalgamation property? For varieties not consisting entirely of completely simple semigroups, we can answer the question fully: such a variety has the weak [strong] amalgamation property if and only if it is a variety of orthodox normal bands of abelian groups. For varieties of completely simple semigroups with abelian subgroups, we can again answer the question fully: all these varieties have the strong amalgamation property. We outline the difficulties presented by the remaining varieties of completely simple semigroups.

1. Introduction

Imaoka [1976a] has determined which varieties of bands have the weak or strong amalgamation property, and Hall [1978a] has answered the corresponding question (modulo a conjecture of group theory) for varieties of inverse semigroups. Given the recent work of Petrich [1977c] on the lattice of varieties of completely regular semigroups, it is natural to consider amalgamation questions for these varieties.

We regard completely regular semigroups as universal algebras with two operations, these being the usual multiplication and a unary operation of inversion (so that x^{-1} denotes the group-inverse of x in the H-class to which x belongs).

For definitions of the amalgamation properties the reader is referred to [Hall, 1978b], and for the structure of completely regular semigroups (and, in particular, normal bands of groups) to [Petrich, 1973].

ISBN 0-12-319450-4

2. Normal bands of groups

The following theorem is due to T.E. Hall (verbal communication, 1979).

THEOREM 1. *If a variety $\mathcal{V}$ of completely regular semigroups has the weak amalgamation property, then it is a variety of normal bands of groups.*

Proof. Let $S \in \mathcal{V}$, and suppose that S is not a normal band of groups. Then there exist (distinct) $\mathcal{J}$-classes S_1, S_2 of S with $S_2 < S_1$, and idempotents $e \in S_1$, $f, g \in S_2$ such that $f < e$, $g < e$ and $f \neq g$ (by [Petrich, 1973]).

If f and g are $\mathcal{R}$-related, then $\{e, f, g\}$ is a right regular, but not a normal, band belonging to $\mathcal{V}$. If f and g are not $\mathcal{R}$-related, put $h = (gf)(gf)^{-1}$. Then $\{e, f, h\}$ is a left regular, but not a normal, band belonging to $\mathcal{V}$.

Thus $\mathcal{V}$ contains either all left regular bands or all right regular bands (since any left [right] regular band which is not normal generates the variety of left [right] regular bands). But the variety of left [right] regular bands, and any larger class of semigroups, does not have the weak amalgamation property (see [Imaoka, 1976a] and remark 7 of [Hall, 1978b]). So, in particular, $\mathcal{V}$ does not have the weak amalgamation property.

THEOREM 2. *Let $\mathcal{V}$ be a variety of normal bands of groups which does not consist entirely of completely simple semigroups, and let $\mathcal{S}$ be the variety of all completely simple semigroups in $\mathcal{V}$. If $\mathcal{V}$ has the weak [strong] amalgamation property, then $\mathcal{S}$ has the congruence extension property and the weak [strong] amalgamation property.*

Proof. Suppose that $\mathcal{S}$ does not have the congruence extension property. Then there is a semigroup S in $\mathcal{S}$ with a subsemigroup U, also in $\mathcal{S}$, having a congruence ρ that does not extend to a congruence on S.

We construct $T = U \dot\cup U/\rho$, a normal band of groups, by using $\phi = \rho^\natural$ as a structure homomorphism. Suppose that the amalgam $(S, T; U)$ can be embedded in a normal band of groups, W, say. Let W_α, W_β be the $\mathcal{J}$-classes of W containing S, $T \setminus U$ respectively, and let $\theta_{\alpha,\beta}$ be the structure homomorphism of W_α into W_β. It can be shown that $\theta_{\alpha,\beta}|U = \phi$, whence $\ker(\theta_{\alpha,\beta}|S)$ extends ρ to S, a contradiction. So S has the congruence extension property. The rest of the theorem is obvious.

REMARK 1. In the case where S is a variety of groups, this result (and the method of proof) is due to Hall [1978a].

THEOREM 3. *Let V be a variety of completely regular semigroups which does not consist entirely of completely simple semigroups. Then V has the weak [strong] amalgamation property if and only if V is a variety of orthodox normal bands of abelian groups.*

Proof. If V has the weak amalgamation property, then V is a variety of normal bands of groups. Let S be the variety of all completely simple semigroups in V. By theorem 2, S has the congruence extension property. From the work of Biró, Kiss and Pálfy [1977a] and Rasin [1979] we can deduce that a variety of completely simple semigroups has the congruence extension property only if it is a variety of rectangular groups with abelian subgroups. If S is such a variety, then V is a variety of orthodox normal bands of abelian groups.

Conversely, let V be a variety of orthodox normal bands of abelian groups. If V is the variety of all orthodox [left, right] normal bands of abelian groups, then Imaoka [1976a] has shown that V has the strong amalgamation property. He obtains this result by observing that the variety of commutative inverse semigroups and the variety of [left, right] normal bands both have the strong amalgamation property, and that orthodox [left, right] normal bands of abelian groups can be characterized as spined products of commutative inverse semigroups and [left, right] normal bands. Since any variety of commutative inverse semigroups has the strong amalgamation property (from [Howie, 1968], [Imaoka, 1976b] and [Hall, 1978a]), and since any variety of normal bands has the strong amalgamation property [Imaoka, 1976a], the method of Imaoka in fact shows that any variety of orthodox normal bands of abelian groups has the strong amalgamation property.

It is now natural to turn our attention to completely simple semigroups.

3. Completely simple semigroups

Imaoka [1977b] has shown that the class of completely 0-simple semigroups has the strong amalgamation property. An easy modification of his proof gives the following theorem.

THEOREM 4. *Let K be a variety of groups, and let CS(K) denote the variety of completely simple semigroups whose subgroups lie in K. Then K has the weak* [*strong*] *amalgamation property if and only if CS(K) has the same property.*

Proof. Let $(S_1,S_2;U)$ be an amalgam in $CS(K)$. Suppose that $S_1 = M(G_1;I_1,\Lambda_1;P_1)$, $S_2 = M(G_2;I_2,\Lambda_2;P_2)$, $U = M(H;J,M;Q)$ and that $P_1 = (p^{(1)}_{\lambda i})$, $P_2 = (p^{(2)}_{\lambda i})$ and $Q = (q_{\lambda i})$. Then G_1,G_2 and H all belong to K.

As in [Imaoka, 1977b] we may assume that $G_1 \cap G_2 = H$, $I_1 \cap I_2 = J$, $\Lambda_1 \cap \Lambda_2 = M$ and $q_{\mu j} = p^{(1)}_{\mu j} = p^{(2)}_{\mu j}$ for all $\mu \in M$, $j \in J$.

Now the amalgam $(G_1,G_2;H)$ can be weakly embedded in a group G, also belonging to K, by monomorphisms ϕ_1 and ϕ_2 of G_1 and G_2 respectively. So $\phi_1|H = \phi_2|H$. Put $I = I_1 \cup I_2$ and $\Lambda = \Lambda_1 \cup \Lambda_2$. Define the $(\Lambda \times I)$-matrix $P = (p_{\lambda i})$ by

$$p_{\lambda i} = \begin{cases} p^{(1)}_{\lambda i}\phi_1 & \text{if } \lambda \in \Lambda_1 \text{ and } i \in I_1, \\ p^{(2)}_{\lambda i}\phi_2 & \text{if } \lambda \in \Lambda_2 \text{ and } i \in I_2, \\ 1 \text{ otherwise}, \end{cases}$$

where 1 denotes the identity element of G. Put $S = M(G;I,\Lambda;P)$. Clearly, $S \in CS(K)$.

Define $\theta_1\colon S_1 \to S$ by $(a;i,\lambda) \to (a\phi_1;i,\lambda)$ and $\theta_2\colon S_2 \to S$ by $(a;i,\lambda) \to (a\phi_2;i,\lambda)$. Then θ_1 and θ_2 are easily seen to be monomorphisms of S_1 and S_2, respectively, into S; and, clearly, θ_1 and θ_2 agree on U. So $(S_1,S_2;U)$ can be weakly embedded in $S \in CS(K)$. If we assume that $(G_1,G_2;H)$ can be strongly embedded in G, then the above construction shows that $(S_1,S_2;U)$ can be strongly embedded in S. The other half of the theorem is obvious.

Rasin [1979] has given a description of all varieties of completely simple semigroups with abelian subgroups. From this we deduce the following theorem.

THEOREM 5. *If V is a variety of completely simple semigroups whose subgroups are all abelian, then V has the strong amalgamation property.*

Proof. It is well known (see, for example, [Neumann, 1967]) that every variety of abelian groups has the strong amalgamation property. In the proof of theorem 4, let K be a variety of abelian groups. Suppose that S_1, S_2 and U all belong to some subvariety V of $CS(K)$. Suppose further that P_1, P_2 and Q are all normalized with respect to some common row and column; that is, let μ_0 and j_0 be elements of M and J respectively, and suppose that every entry in the μ_0-row and the j_0-column of P_1, P_2 and Q is 1. Then the semigroup S, constructed in the proof of Theorem 4, also belongs to V. Thus V has the strong amalgamation property.

It is easy to show that a variety of rectangular groups has the weak [strong] amalgamation property if and only if the corresponding variety of groups has the same property. Now if a variety K of groups has the weak amalgamation property, and if K is not the variety of all groups, then every finite group in K is abelian (see [Hall, 1978a], where the result is attributed to B.H. Neumann; see also [Neumann, 1967]). We can now state the following theorem.

THEOREM 6. *The following varieties of completely simple semigroups have the strong amalgamation property:*

(a) the varieties of all groups, left groups, right groups, rectangular groups and completely simple semigroups, and

(b) all varieties of completely simple semigroups with abelian subgroups.

Conversely, any variety of completely simple semigroups with the weak [strong] amalgamation property is one of the above or is

(c) a non-abelian variety K of groups with all its finite members abelian, the variety $CS(K)$ of all completely simple semigroups whose subgroups lie in K, or a subvariety of $CS(K)$ containing K, or

(d) a variety of completely simple semigroups containing the variety of rectangular groups.

REMARK 2. If there exists a variety K of groups as in (c), and having the weak [strong] amalgamation property, then the variety $CS(K)$ and the corresponding varieties of left, right and rectangular groups all have the same property. We cannot answer whether or not any of the other varieties as in (c) have the weak [strong] amalgamation property. Nor can we answer whether or not any varieties as in (d) (other than the variety of completely simple semigroups and the variety of rectangular groups) have the weak or strong amalgamation property.

Acknowledgements

I wish to thank my supervisor, Dr T.E. Hall, for much valuable help and advice. I also wish to thank Dr P.R. Jones for a number of useful suggestions.

The support of a Monash Graduate Scholarship is gratefully acknowledged.

References

1967 H. Neumann, *Varieties of groups*. Springer-Verlag, Berlin, Heidelberg, 1967.

1968 J.M. Howie, Commutative semigroup amalgams. *J. Austral. Math. Soc.*, 8 (1968), 609-30.

1973 M. Petrich, *Introduction to semigroups*. Merrill, Columbus, 1973.

1976a T. Imaoka, Free products with amalgamation of bands. *Mem. Fac. Lit. & Sci., Shimane Univ. Nat. Sci.*, 10 (1976), 7-17.

1976b T. Imaoka, Free products with amalgamation of commutative inverse semigroups. *J. Austral. Math. Soc.*, 22 (A) (1976), 246-51.

1977a B. Biró, E.W. Kiss and P.P. Pálfy, On the congruence extension property. *Semigroup Forum*, 15 (1977), 183-4.

1977b T. Imaoka, Free products with amalgamation of semigroups. Ph.D. Thesis, Monash University, 1977.

1977c M. Petrich, Certain varieties and quasivarieties of completely regular semigroups. *Can. J. Math.*, 29 (1977), 1171-97.

1978a T.E. Hall, Inverse semigroup varieties with the amalgamation property. *Semigroup Forum*, 16 (1978), 37-51.

1978b T.E. Hall, Representation extension and amalgamation for semigroups. *Quart. J. Math.* Oxford (2), 29 (1978), 309-34.

1979 V.V. Rasin, On the lattice of varieties of completely simple semigroups. *Semigroup Forum*, 17 (1979), 113-22.

EMBEDDING THEOREMS USING AMALGAMATION BASES

C. J. Ash

Department of Mathematics, Monash University, Clayton,
Victoria, Australia, 3168.

Abstract

We consider some applications of the general theory of homogeneous structures to classes of semigroups. The usual construction for these requires the amalgamation property, but we give a variation which applies to many more classes, involving the notion of an amalgamation base. We thus obtain analogues of the Higman, Neumann and Neumann Embedding Theorem for the class of inverse semigroups and for the class of all semigroups.

We also obtain a widely applicable theorem embedding a semigroup in a given class into one in the same class which is "strongly uniform" with respect to its idempotents. We illustrate a use of this by proving that every orthodox semigroup may be embedded in a bisimple orthodox semigroup.

We examine here some results of the form that a semigroup S may be embedded into a semigroup H so that H has automorphisms extending certain isomorphisms between subsemigroups of S. Such questions may also be asked when the semigroups under consideration are restricted to belong to a fixed class of semigroups (bands, regular semigroups, etc.).

We consider an arbitrary class of semigroups which is closed under isomorphism and under the formation of unions of chains. A long-standing result of B. Jónsson [1960] gives a result of the kind mentioned for such classes also having the *amalgamation property*. There is a similar result, involving the notion of an *amalgamation base* for such classes which do not necessarily have the amalgamation property but which are closed under the formation of *ultraproducts*. We give some applications of these results.

ISBN 0-12-319450-4

1. Classes with the amalgamation property

Let M be a class of semigroups satisfying the following two conditions.

(I) M is closed under isomorphism.

(II) M is closed under formation of unions of chains.

Let κ be a cardinal number.

DEFINITION. A semigroup $H \in M$ is *(M,κ)-homogeneous* if for all $U,V \leqslant H$ with $U,V \in M$ and $|U|,|V| < \kappa$ and for all $f\colon U \simeq V$, there exists an automorphism of H extending f.

DEFINITION. The class M has the *amalgamation property* (AP) if for all $U,S,T \in M$ and all monomorphisms $f\colon U \to S$, $g\colon U \to T$ there exist $A \in M$ and monomorphisms $h\colon S \to A$, $k\colon T \to A$ for which the following diagram commutes.

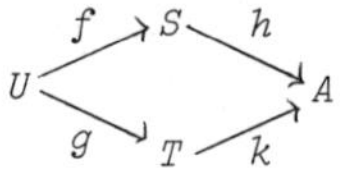

THEOREM 1.1 [Jónsson 1960]. *Let M be a class of semigroups satisfying conditions (I) and (II) and having (AP). Then for every $S \in M$ and every cardinal κ, there exists $H \in M$ for which H is (M,κ)-homogeneous and for which $S \leqslant H$.*

2. Amalgamation bases

DEFINITION. A semigroup $U \in M$ is an *amalgamation base* for M if for all $S,T \in M$ and all monomorphisms $f\colon U \to S$, $g\colon U \to T$, there exist $A \in M$ and monomorphisms $h\colon S \to A$, $k\colon T \to A$ for which the following diagram commutes.

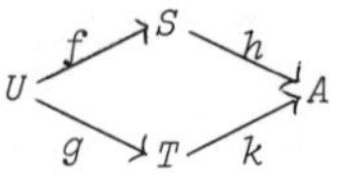

Thus, M has (AP) if and only if *every* $U \in M$ is an amalgamation base for M. It does not seem an easy matter to determine, in a given class M of semigroups, whether a particular semigroup is an amalgamation base, but a fairly general condition on M ensures that there is a "large supply" of amalgamation bases.

We consider the following further condition on the class M.

(U) M is closed under the formation of ultraproducts.

The definition of the ultraproduct construction may be found in [Grätzer 1968]. A fuller discussion appears in [Bell and Slomson 1969]. Many commonly occurring classes of general algebras satisfy condition (U), notably all those known in model theory as "elementary classes". These include all varieties and quasi-varieties and many other classes such as those mentioned in section 3.

THEOREM 2.1. *Let* M *be a class of semigroups satisfying conditions (I), (II) and (U).*

(i) The class of amalgamation bases for M *is closed under the formation of unions of chains.*

(ii) Every $S \in M$ *may be embedded in an amalgamation base for* M.

DEFINITION. A semigroup $H \in M$ is *(M,κ)-homogeneous for amalgamation bases* if for all $U, V \leqslant H$ with U, V amalgamation bases for M and $|U|, |V| < \kappa$ and for all $f\colon U \simeq V$, there exists an automorphism of H extending f.

THEOREM 2.2. *Let* M *be a class of semigroups satisfying conditions (I), (II) and (U). Then for every* $S \in M$ *and every cardinal* κ*, there exists* $H \in M$ *for which* H *is* (M,κ)*-homogeneous for amalgamation bases and for which* $S \leqslant H$.

Proofs of theorems 2.1 and 2.2 are given by the author [B].

3. Remarks on classes and amalgamation bases

Some examples

The following classes of semigroups all satisfy conditions (I), (II) and (U). Theorem 2.1 therefore applies. In those cases where (AP) does not hold, there are listed those easily described amalgamation bases known to the author.

CLASS	AMALGAMATION BASES
All semigroups	Inverse semigroups [Howie 1975]. Finite cyclic semigroups, two-element left [right] zero bands [Hall 1978b]
Regular semigroups	Inverse semigroups, two-element left [right] zero bands [theorem 3.3]
Groups	(AP) [Schreier 1927]
Inverse semigroups	(AP) [Hall 1975]
Semilattices	(AP) [Howie 1968]
Commutative groups	(AP) [Neumann 1954]
Commutative inverse semigroups	(AP) [Howie 1968]
Commutative semigroups	Finite cyclic semigroups, commutative inverse semigroups, commutative totally division ordered semigroups [Howie 1968]
Bands	Semilattices which form distributive lattices [corollary 3.2]
Orthodox semigroups	One-element semigroups [lemma 5.4]

By contrast, the class of locally finite semigroups does not satisfy condition (U) and the class of finite semigroups satisfies neither (II) nor (U). We do not know in these and similar cases whether theorem 2.1 applies.

Sources of amalgamation bases

The only general methods known to the author for obtaining particular amalgamation bases are by forming unions of chains, using theorem 2.1(i), and by use of the following

THEOREM 3.1. *Let* M *be a class of semigroups closed under the formation of finite direct products. Then every injective member of* M *is an amalgamation base for* M.

Proof. Let $U \in M$ be injective in M. Then in particular, for every $S \in M$ and every monomorphism $f\colon U \to S$ there exists a homomorphism $\phi\colon S \to U$ such that, for all $u \in U$, $\phi(f(u)) = u$.

Now suppose that $S, T \in M$ and $f\colon U \to S$, $g\colon U \to T$ are monomorphisms. Then there exist homomorphisms $\phi\colon S \to U$, $\psi\colon T \to U$ such that, for all $u \in U$, $\phi(f(u)) = u = \psi(g(u))$. Define $h\colon S \to S \times T$ by $h(s) = \langle s, g(\phi(s))\rangle$ and $k\colon T \to S \times T$ by $k(t) = \langle f(\psi(t)), t\rangle$. Then h, k are monomorphisms and, for all $u \in U$, $h(f(u)) = \langle f(u), g(\phi(f(u)))\rangle = \langle f(u), g(u)\rangle = \langle f(\psi(g(u))), g(u)\rangle = k(g(u))$. Thus the following diagram commutes.

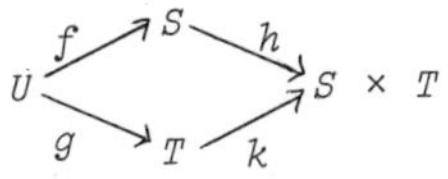

Hence U is an amalgamation base for M.

COROLLARY 3.2. *In any variety of bands, every semilattice which forms a distributive lattice is an amalgamation base.*

Proof. From [Horn and Kimura 1971], every finite semilattice which forms a distributive lattice is injective in the class of semilattices and so, by [Gerhard 1974], is injective in every variety of bands. By theorem 3.1, such semilattices are amalgamation bases. The result follows since, by theorem 2.1(i), amalgamation bases are closed under unions of chains and therefore under unions of directed systems.

COMMENT. It does not appear to be known whether or not *every* semilattice is an amalgamation base in the class of all bands.

A final comment about amalgamation bases concerns the following notion.

DEFINITION. Let M_0, M be classes of semigroups for which $M_0 \subseteq M$. The class M_0 is said to be *cofinal* in M if for all $S \in M$ there exists $T \in M_0$ with $S \leqslant T$.

Thus, the class of regular semigroups is cofinal in the class of all semigroups. The following theorem is immediate from the definitions.

THEOREM 3.3. *If M_0 is cofinal in M then $S \in M_0$ is an amalgamation base for M_0 if and only if it is for M.*

4. Embedding theorems using the amalgamation property

By theorem 1.1, for every κ, any inverse semigroup may be embedded into a κ-homogeneous inverse semigroup. We have the following modifications of this.

THEOREM 4.1. *Let S be an inverse semigroup.*

(i) S may be embedded in an inverse semigroup T with 1 such that every isomorphism between inverse subsemigroups of S extends to an inner automorphism of the form $s \mapsto g^{-1}sg$ for some $g \in T$ with $g^{-1}g = gg^{-1} = 1$.

(ii) For every κ, S may be embedded in a κ-homogeneous inverse semigroup H such that every isomorphism between inverse subsemigroups of H of cardinality less than κ is of the form $s \mapsto g^{-1}sg$ for some group element g of H.

Proof. (i) Let $\kappa_0 > |S|$. Let $H \geqslant S$ be a κ_0-homogeneous inverse semigroup (theorem 1.1). Embed H into $I(H^1) = T$.

(ii) Let $\kappa_0 > |S|$, $\kappa_0 \geqslant \kappa$. Let $S \leqslant H_1 \leqslant T_1$ be as in the proof of (i). Let $\kappa_1 > |T_1|$. Take $T_1 \leqslant H_2 \leqslant T_2$ as for (i). Continuing in this way, taking unions at limit ordinals, we obtain a chain $\{T_\xi\}$ of length κ^+ whose union has the desired property.

COMMENTS. A κ-homogeneous inverse semigroup (for $\kappa > 1$) which is not a group cannot have 1. So in theorem 4.1, unless S is a group we cannot have $T = H$.

Theorem 4.1 thus provides two possible generalizations of the well-known embedding theorem of Higman, Neumann and Neumann [1949] for groups. In particular, two elements of an inverse semigroup generating isomorphic inverse subsemigroups can be made conjugate in either of these senses.

Our theorem 4.1(i) is also given in [A].

DEFINITION. Say that a semilattice, E, is *strongly uniform* if for all $e,f \in E$ there is an automorphism of E with $e \mapsto f$.

NOTE. If E is a strongly uniform semilattice then, for all $e,f \in E$ we have $Ee \simeq Ef$. The converse is not true in general.

Clearly a κ-homogeneous semilattice (for $\kappa > 1$) is strongly uniform. So theorem 1.1, applied to the class of all semilattices, gives the following.

THEOREM 4.2. *Every semilattice may be embedded in a strongly uniform semilattice.*

5. Embedding theorems using amalgamation bases

Using theorem 2.2 instead of theorem 1.1, we may prove the following theorem, which is similar to theorem 4.1.

THEOREM 5.1. *Let S be a semigroup.*

(i) S may be embedded in a semigroup T with 1 such that every isomorphism between subsemigroups of S which are amalgamation bases in the class of semigroups extends to an inner automorphism of the form $s \mapsto g^{-1}sg$ for some $g \in T$ with $g^{-1}g = gg^{-1} = 1$.

(ii) For every κ, S may be embedded in a semigroup H, κ-homogeneous for amalgamation bases, such that every isomorphism between subsemigroups of H of cardinality less than κ which are amalgamation bases in the class of semigroups is of the form $s \mapsto g^{-1}sg$ for some group element g of H.

COMMENT. Thus, in either of these senses, elements of S may be made conjugate if they generate isomorphic *finite* subsemigroups, since finite cyclic semigroups are amalgamation bases, but not necessarily if the elements generate infinite subsemigroups.

DEFINITION. Say that a band, B, is *strongly uniform* if for all $e,f \in B$ there exists an automorphism of B with $e \mapsto f$.

LEMMA 5.2. *The 1-element semigroup is an amalgamation base in the class of bands.*

Proof. Take $f: 1 \to B_1$, $g: 1 \to B_2$ where B_1, B_2 are bands. Let $f(1) = e_1$, $g(1) = e_2$. Define $h: B_1 \to B_1 \times B_2$ and $k: B_2 \to B_1 \times B_2$ by $b_1 \mapsto (b_1, e_2)$ and $b_2 \mapsto (e_1, b_2)$ respectively. Then h,k are monomorphisms and $h(f(1)) = (e_1, e_2) = k(g(1))$, so the following diagram commutes.

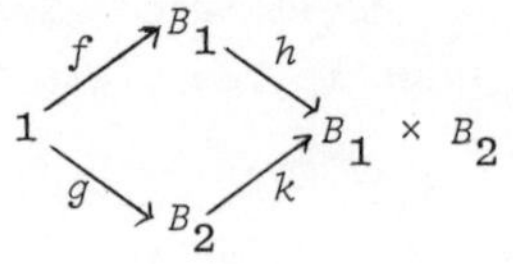

THEOREM 5.3. *Every band may be embedded in a strongly uniform band.*

Proof. The theorem follows from lemma 5.2 and theorem 2.2.

Clearly, using the same method of proof, we have the following.

THEOREM 5.4. *Let* M *be any class of semigroups containing a 1-element semigroup, satisfying conditions (I), (II) and (U) and closed under the formation of finite direct products.*

(i) The 1-element semigroup is an amalgamation base for M.

(ii) For every $S \in M$ *there exists* $T \in M$ *with* $S \leqslant T$ *and such that for any two idempotents* $e, f \in T$*, there is an automorphism of* T *with* $e \to f$.

We conclude with an application, due to T.E. Hall, of this theorem to orthodox semigroups.

THEOREM 5.5. *Every orthodox semigroup may be embedded into a bisimple orthodox semigroup.*

Proof. Let S be orthodox. Then, by theorem 5.4, S may be embedded into an orthodox semigroup H whose band, B, is strongly uniform. By the construction of [Hall 1970] there is a bisimple orthodox semigroup $W(B)$ with band B. Now by [Hall 1978a, theorem 8 and corollary 9], H and $W(B)$ may be amalgamated over B into an orthodox semigroup V with band B. Thus any two idempotents of V are in B, hence are $\mathcal{D}$-related in $W(B)$ and therefore in V. So V is bisimple.

CONCLUSION. Although, under the assumptions of theorem 2.1, a class M will have a large number of amalgamation bases, there is no *a priori* reason why any of these should be finitely generated, let alone finite. We hope to have shown how, when simply described amalgamation bases are found, they can be put to use.

Even in the case where M is a variety, few general properties of the class A(M) of amalgamation bases for M seem to be known. From this point of view it would be useful to obtain simple characterizations of A(M) for the commonly occurring classes M of semigroups, if such characterizations are indeed possible.

ACKNOWLEDGEMENT. The author wishes, as always to thank Tom Hall for his stimulating conversations and, on this occasion, for his proof of theorem 5.5. Thanks also go to H.E. Scheiblich for the information yielding corollary 3.2.

References

1927 O. Schreier, Die untergruppen der freien gruppen. *Abh. Math. Sem. Univ. Hamburg* 5 (1927), 161-183.

1949 G. Higman, B. Neumann and H. Neumann, Embedding theorems for groups. *J. Lond. Math. Soc.* 24 (1949), 247-254.

1954 B.H. Neumann, An essay on free products of groups with amalgamation, *Philos. Trans. Roy. Soc. London Ser. A* 246 (1954), 503-554.

1960 B. Jónsson, Homogeneous universal relational systems. *Math. Scand.* 8 (1960), 137-142.

1968 G. Grätzer, *Universal Algebra*, Van Nostrand, New York, 1968.

1968 J.M. Howie, Commutative semigroup amalgams. *J. Aust. Math. Soc.* 8 (1968), 609-630.

1969 J. Bell and A. Slomson, *Models and ultraproducts*, North Holland, Amsterdam, 1969.

1970 T.E. Hall, On orthodox semigroups and uniform and antiuniform bands, *J. Algebra* 16 (1970), 204-217.

1971 A. Horn and N. Kimura, The category of semilattices. *Algebra Universalis* 1 (1971) 26-38.

1974 J.A. Gerhard, Injectives in equational classes of idempotent semigroups. *Semigroup Forum* 9 (1974) 36-53.

1975 T.E. Hall, Free products with amalgamation of inverse semigroups. *J. Algebra* 34 (1975), 375-385.

1975 J.M. Howie, Semigroup amalgams whose cores are inverse semigroups. *Quart. J. Math. Oxford (2)* 26 (1975) 23-45.

1978a T.E. Hall, Amalgamation and inverse and regular semigroups. *Trans. Amer. Math. Soc.* 246 (1978), 395-406.

1978b T.E. Hall, Representation extension and amalgamation for semigroups. *Quart. J. Math. Oxford (2)* 29 (1978), 309-334.

TO BE PUBLISHED

[A] T.E. Hall, Inverse and regular semigroups and amalgamation. *Proceedings of the Conference on regular semigroups, De Kalb,* April 1979 (to appear).

[B] C.J. Ash, Structures homogeneous for amalgamation bases (submitted for publication).

NOTE ADDED IN PROOF

Theorems 2.1 and 2.2 have been previously proved by M. Yasuhara, The amalgamation property, the universal-homogeneous models, and the generic models, *Math. Scand.* 34 (1974), 5-36. Theorem 2.1 comprises his proposition 1.6.2 and theorem 1.7, while our theorem 2.2 follows from his theorem 2.5 using his earlier remarks and the assumption on κ. A slightly different proof establishes our theorem 2.2 without any assumption on κ.

PARTIALLY ORDERED SEMIGROUPS AS SEMIGROUPS

R. McFadden

Department of Mathematical Sciences, Northern Illinois University,
De Kalb, Illinois, U.S.A., 60115.

Abstract

The purely algebraic theory of semigroups and the theory of residuated semigroups as developed by the French school have had very little overlap until recently. This expository article is intended to draw attention to the positive gains that have been made by utilizing the algebraic structure of a residuated semigroup to obtain more information about its structure as a whole. This approach has been fruitful in terms both of the results obtained and the methods used. Outlines of proofs have been included where the interplay of algebraic and ordering ideas are particularly striking.

This paper contains a survey of some recent publications on partially ordered semigroups in which the authors have used the algebraic structure of the semigroup to obtain new and illuminating results about its overall structure. As such, the paper will not be concerned with such results as, for example, those of Saitô [1971a], McAlister [B], or Behrens [1972a, A] which begin with semigroups of a given algebraic structure, say inverse semigroups or completely simple semigroups, and proceed to investigate the inter-relation between the ordering and the algebraic structure. Nor will the paper be concerned solely with such particular types of partially ordered semigroups as, for example, integrally closed residuated semigroups, in which the dominant theme is that of the ordering [Molinaro, 1960, 1961a].

1. Residuated semigroups

Consider a partially ordered semigroup $S(\leqslant,\cdot)$, that is a multiplicative semigroup $S(\cdot)$ which is also a partially ordered set $S(\leqslant)$ in which the multiplication is compatible with the ordering. It may happen

ISBN 0-12-319450-4

that for arbitrary elements a,b of S, the sets

$$\{x \in S \mid ax \leq b\} \quad \text{and} \quad \{x \in S \mid xa \leq b\}$$

are non-empty. Such is the case when S is the set of subsets of a semigroup T, and S is partially ordered by inclusion. It is also the case when S is an inverse semigroup and $\leq$ is the natural partial ordering on S; one simply takes $x = a^{-1}b$ or $x = ba^{-1}$, respectively. In fact, if S is an E-unitary inverse semigroup with subsemigroup E of idempotents, then $E = \{x \in S \mid x^2 \leq x\}$; if S has an identity element 1, say, then

$$ax \leq 1 \text{ if and only if } xa \leq 1,$$

and the minimum group congruence σ on S may be described by

$$(a,b) \in \sigma \text{ if and only if } ax \leq 1 \text{ is equivalent to } bx \leq 1.$$

For comprehensive treatments of the algebraic theory of semigroups see Clifford and Preston [1961b] and Howie [1976].

In general, if for a pair of elements a,b of a partially ordered semigroup S the set $\{x \in S \mid ax \leq b\}$ contains a maximum element, that element is called the right residual of b by a, and we shall denote it here by $a^{[-1]}b$. This notation is not new; it has been used, for example, in Clifford and Preston [1961b] in a set-theoretic sense, and a similar one was used in Eilenberg [1974d] in the development of a division calculus. It is used here deliberately to suggest an analogue to the inverse of an element in an inverse semigroup, though of course $a^{[-1]}$ by itself is not defined. There is a left dual definition of $ba^{[-1]}$, and if both elements exist for all a,b in S then S is said to be a residuated semigroup. For a comprehensive treatment of residuated semigroups see Blyth and Janowitz [1972b]. As an example of a residuated semigroup, one may note that the ω-bisimple inverse semigroup $B(G,\alpha)$ is residuated under its natural partial ordering if and only if α is an automorphism [1971b].

Some of the more common rules for calculating with residuals are the following: if $a,b,c \in S$ then

$$a \leq b \text{ implies } \begin{cases} c^{[-1]}a \leq c^{[-1]}b \text{ and } ca^{[-1]} \geq cb^{[-1]} \\ ac^{[-1]} \leq bc^{[-1]} \text{ and } a^{[-1]}c \geq b^{[-1]}c. \end{cases}$$

$(b^{[-1]}a)c^{[-1]} = b^{[-1]}(ac^{[-1]})$ is the maximum x in S satisfying $bxc \leqslant a$.

For $a = b = c$ it is tempting to write $a^{-1} = a^{[-1]}aa^{[-1]}$, but this is the wrong analogy to the inverse case.

I. Molinaro [1960, 1961a] defined two equivalence relations on an arbitrary residuated semigroup S which have been of crucial importance in the theory of residuated semigroups in general. They are defined in terms of a fixed element n of S as follows:

$$(a,b) \in {}_nA \quad \text{if and only if} \quad na^{[-1]} = nb^{[-1]}$$

and

$$(a,b) \in {}_nB \quad \text{if and only if} \quad an^{[-1]} = bn^{[-1]}.$$

The relation ${}_nA$ (resp. ${}_nB$) may be characterized as an equivalence relation with convex classes such that each ${}_nA$-class (resp. ${}_nB$-class) contains a maximum (resp. minimum) element. The maximum (resp. minimum) element in the ${}_nA$-class (resp. ${}_nB$-class) of $a \in S$ is $(na^{[-1]})^{[-1]}n$ (resp. $(an^{[-1]})n$). There are right dual definitions and characterizations for A_n and B_n.

Much of the theory of residuated semigroups has centered around the properties of these equivalence relations - their compatibility with multiplication, with residuation, with meet or with join if these exist, the form of the distinguished elements in the equivalence classes. Both the A-type and the B-type equivalences observe some compatibility laws; thus A_n is compatible on the right with multiplication on the right, and the elements n for which any A-type or B-type equivalence is a congruence relation must satisfy stringent conditions. To sum up the situation for an A-type congruence we may use the following theorem [Blyth and Janowitz 1972b].

THEOREM 1. *Let S be a residuated semigroup and let ρ be a congruence on S such that S/ρ is a partially ordered group, the isotone homomorphic image of S under the canonical mapping from S onto S/ρ. Then ρ is a congruence of A-type if and only if one (and hence each) of its classes contains a maximum element. In this case $\rho = {}_mA = A_m = \mathsf{A}$, say, where $m^2 \leqslant m$ and $m \geqslant a^{[-1]}a$, $m \geqslant aa^{[-1]}$ for all a in S. The partially ordered group S/A is the maximum isotone group homomorphic image of S. The congruence A is the unique A-type congruence for which the quotient is an isotone group homomorphic image of S, and m is the maxi-*

mum element of the identity class of $\mathcal{A}$, *which is called the* $\mathcal{A}$*-nomal congruence on* S.

A result like this depends very heavily on the single element m. For example, there is available a ready cross-section of the $\mathcal{A}$-classes, namely the residuals of m. On the other hand, the theorem makes no mention of the algebraic structure of S, and this omission was typical of such theorems until quite recently. Consider another illustration, the case in which ρ is as in theorem 1 and at the same time is an equivalence of B-type, say $\rho = \mathcal{A} = \mathcal{B}$. When S contains an element x for which

$$ {}_xB = B_x = B_{yx} = {}_{xy}B \quad \text{for each } y \text{ in } S, $$

then x is called a B-nomaloid element, $\mathcal{B} = B_x$ is called the (unique) B-nomal congruence on S, and S itself is called a B-nomal semigroup. The element x is called B-nomal if it is minimum in its $\mathcal{B}$-class. When in addition $\mathcal{A} = \mathcal{B}$, the semigroup S is said to be B-nomally closed. McAlister and O'Carroll [1970] characterized B-nomal semigroups in terms of their semigroup structure as in theorem 2.

THEOREM 2. *Let* S *be a residuated semigroup.*

(i) S *is* B*-nomal if and only if* S *has a group kernel,* K, *say.*

(ii) S *is* B*-nomally closed if and only if the identity element* e *of* K *satisfies* $ae \leqslant a$ *for all* a *in* S.

(iii) The following conditions on S *are equivalent:*

(a) S *contains a minimal element.*

(b) S *is* B*-nomally closed and* K *is trivially ordered.*

(c) S *is* B*-nomal and* K *is trivially ordered.*

(d) $K = S\mathcal{B}$, *the set of elements minimum in their* $\mathcal{B}$*-classes.*

There are two main ingredients in the proof of this theorem. The first is that for x in S the set of elements minimum in their ${}_xB$-classes is Sx, and therefore that x is B-nomaloid if and only if

$$ yxS = xS = Sx = Sxy \text{ for each } y \text{ in } S. $$

The second is the observation that Sx is then the minimum left ideal of

S and also the minimum right ideal of S; it is therefore the kernel of S.

Theorem 2 both clarifies and simplifies much of the theory of B-nomal semigroups. It may be useful here to consider an example of a B-nomally closed semigroup; although of simple construction, this example illustrates several of the main features of B-nomal semigroups in general.

EXAMPLE 1. Let S denote the Cartesian product of the circle group and the closed unit interval, with the interval ordered in the usual way and multiplication defined by $ab = \min\{a,b\}$. With $(g,a) \leqslant (h,b)$ if and only if $g = h$ and $a \leqslant b$, it is clear that S is residuated, with

$$(g,a)^{[-1]}(k,c) = \begin{cases} (g^{-1}k, 1) & \text{if } a \leqslant c \\ (g^{-1}k, c) & \text{if } a > c. \end{cases}$$

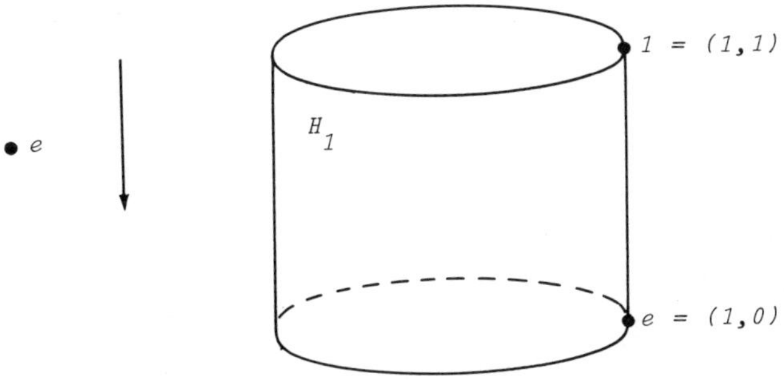

$K = eS$

Let us return to inverse semigroups for a moment. If each σ-class of an inverse semigroup S contains a maximum element with respect to the natural partial ordering (let us denote the set of such elements by T), then S is called an F-inverse semigroup. The structure of an arbitrary F-inverse semigroup was determined by McFadden and O'Carroll [1971b] in terms of the semilattice E of idempotents of S, the maximum group homomorphic image G of S, and a set of endomorphisms of E, one for each element of G. In brief, S is isomorphic to

$$\{(f,t) \in E \times T \mid f \leqslant tt^{-1}\}$$

with multiplication defined by

$$(f,t)(g,u) = (f \wedge t.g, tu),$$

where we have identified t and u with their canonical images in G. It was shown that S is residuated if and only if E is residuated. It proved possible to use a similar technique to determine the structure of, not all, but a certain class of A-nomal semigroups. From theorem 2, it follows that when a residuated semigroup S is B-nomally closed, its group kernel K is the retract of S under the anticlosure mapping associated with B. However, an A-nomal semigroup need not be B-nomal, as may be seen by omitting zero from the unit interval in example 1. To find a result analogous to that of theorem 2, it was necessary first to characterize those A-nomal semigroups for which the maximal subgroup H_e of a residuated semigroup S is the retract of S under the closure mapping associated with ${}_eA$. Since it was known which residuated semigroups contain a maximal element, it was then possible to prove [McFadden, 1973]:

THEOREM 3. *Let S be a residuated semigroup containing a maximal element, let S have an identity element 1, and suppose that*

$$1x^{[-1]} \in H_1, \; x^{[-1]}1 \in H_1 \text{ for each } x \in S.$$

Then the structure of S may be characterized in terms of H_1, the A-*class E of 1, and a set of endomorphisms of E, one for each element of H_1.*

There are examples of semigroups S which satisfy the hypotheses of theorem 3 but are not inverse semigroups even though E may be a lattice. However, one can show that such a semigroup is an F-inverse semigroup, and its ordering is the natural partial ordering, exactly when E is a band. In this case the (inverse semigroup) inverse of an element a of S is given by $a^{-1} = (1a^{[-1]})a(1a^{[-1]})$. These results were later extended by Blyth [1974b] to the case where the set of A-nomal elements form a filter, and to the case where H_1 need not be trivially ordered. The concept of an F-inverse semigroup has also been generalized [Batbedat 1978a] to include the action of one monoid, which need not be a group, on another.

2. Dubreil-Jacotin semigroups

It is not necessary to assume that all residuals exist in order to obtain interesting algebraic results. Consider a Dubreil-Jacotin semigroup, defined as follows.

DEFINITION. A Dubreil-Jacotin semigroup is a partially ordered semigroup S which admits an isotone epimorphism f onto a partially ordered group G such that the inverse image under f of the negative cone of G contains a maximum element m, say, of S. Then m is the maximum element of S which satisfies $a^2 \leqslant a$, and $ab \leqslant m$ if and only if $ba \leqslant m$ for each $a,b \in S$. If $ma^{[-1]}$ exists for each a in S then S is said to be a strong Dubreil-Jacotin semigroup. (See Blyth and Janowitz, [1972b] for an account of Dubreil-Jacotin semigroups.)

As noted above, every E-unitary inverse semigroup with identity is, under its natural partial ordering, a Dubreil-Jacotin semigroup.

Now consider what happens when a Dubreil-Jacotin semigroup S is regular. In fact the element m is then the maximum idempotent of S. But this provides very little more information about the structure of S, for if we take an arbitrary partially ordered set X with maximum element m, say, and let S be the left zero semigroup on X, then S is a regular (strong) Dubreil-Jacotin semigroup. The situation is much better when m is the identity element. This occurs for a regular Dubreil-Jacotin semigroup S exactly when

$$m = x^{[-1]}x = xx^{[-1]} \text{ for each } x \in S.$$

Let us assume from now on that this is the case; we shall denote m by 1 and call S an integrally closed Dubreil-Jacotin semigroup. Note, however, that we do not assume that $1a^{[-1]}$ $(=a^{[-1]}1)$ exists for each a in S. We do have the following result.

THEOREM 4. *Every regular integrally closed Dubreil-Jacotin semigroup is an E-unitary inverse semigroup.*

This was proved by McFadden [1975] using an argument based on a combination of algebraic and ordering techniques. One shows first that when restricted to the set E of idempotents the partial ordering of the semigroup S is the natural partial ordering of the idempotents of a regular semigroup. It follows that E is the identity class of A, and is therefore a partially ordered semigroup with maximum element 1, and hence is a band. On the inverse semigroup S one then shows that $A \subseteq \sigma$, so the identity class of σ is exactly E, as required.

With the hypothesis of theorem 4 the natural partial ordering coincides with $\leqslant$ on every σ-class, not just on E. But although S/σ and S/A are algebraically isomorphic they are not in general isomorphic as partially ordered groups.

EXAMPLE 2. Let S be the additive group Z of the integers, and let $\leqslant$ be the usual ordering on Z. Then S is a regular (strong) integrally closed Dubreil-Jacotin semigroup on which σ and A are both equality, so S/σ and S/A are algebraically isomorphic as groups; but S/σ is trivially ordered and S/A is isomorphic to Z with the usual ordering.

Let us now turn to the problem of classifying the partial orderings on an integrally closed Dubreil-Jacotin inverse semigroup S. So far this has not been done in general, but only for those orderings for which Green's relation R is regular in the sense introduced by Blyth [1974a]

$$x \leqslant y \text{ implies } xx^{-1} \leqslant yy^{-1} \text{ for } x,y \in S.$$

This is equivalent to saying that the canonical mapping from S onto S/R, ordered by $xR \leqslant yR$ if and only if $xx^{-1} \leqslant yy^{-1}$, is isotone. One can define a left dual notion of regularity for h, though the two conditions are mutually independent.

The E-unitary inverse semigroups are exactly those in which $\sigma \cap R$ is equality. Using this, McAlister [1974c] has determined the structure of all E-unitary inverse semigroups; they are precisely the inverse semigroups $P = P(G,X,Y)$, where X is a partially ordered set on which the group G acts by order isomorphisms, and Y is a subsemilattice ideal of X. Then $P = \{(a,g) \in Y \times G | g^{-1}a \in Y\}$ under the product

$$(a,g)(b,h) = (a \wedge gb, gh),$$

and P has an identity element if and only if Y has a maximum element. After theorem 4, we may assume that every integrally closed Dubreil-Jacotin inverse semigroup is a P-semigroup. The main result in the direction of classifying our orderings on P is [McFadden 1975].

THEOREM 5. *Let $P = P(G,X,Y)$ be an E-unitary inverse semigroup* *with identity element. There is a one-to-one correspondence between partial orderings on P for which* R *is regular and P is an integrally closed Dubreil-Jacotin inverse semigroup, and partial orderings* $\leqslant$ *on G for which G is a partially ordered group satisfying*

$$(a,g),(b,h) \in P,\ a \leqslant b \text{ in } Y,\ g \leqslant h \text{ in } G$$

imply

$$a \wedge gc \leqslant b \wedge hc \quad \text{in } X \text{ for each } c \in Y.$$

One proves this by noting first that since G is isomorphic to P/σ, it follows from the proof of theorem 4 that G is isomorphic to P/A; one can therefore define a partial ordering $\leqslant$ on G by using that on the partially ordered group P/A. One then uses the regularity of R to establish the necessity and sufficiency of the condition stated in the theorem. In the course of the proof it becomes clear that P must be ordered coordinatewise.

There is a corresponding theorem for the case in which L is regular; the relevant condition to be satisfied is:

$$(a,g),(b,h) \in P,\ a \leqslant b \text{ in } Y,\ g \leqslant h \text{ in } G$$

imply

$$g^{-1}(a \wedge c) \leqslant h^{-1}(b \wedge c) \text{ in } X \text{ for each } c \in Y.$$

The conditions of theorem 5 and of its left dual are considerably simplified when P is an F-inverse semigroup. This is precisely the case when, calculating via the natural partial ordering on P, $(1,1)(a,g)^{[-1]} = (a,g)^{[-1]}(1,1)$ exists for each (a,g) in P. We find then that P is a strong Dubreil-Jacotin inverse semigroup if and only if P is F-inverse and, calculating now via the ordering $\leqslant$,

$$(a,g) \leqslant (b,h) \text{ implies } (1 \wedge g1,1) \leqslant (1 \wedge h1,1) \text{ in } P.$$

To find which E-unitary inverse semigroups are strong integrally closed Dubreil-Jacotin inverse semigroups, we may therefore concentrate on F-inverse semigroups and obtain [McFadden 1975] :

THEOREM 6. *Let $P = P(G,X,Y)$ be an F-inverse semigroup. There is a one-to-one correspondence between partial orderings on P for which* R *is regular and P is a strong integrally closed Dubreil-Jacotin inverse* *semigroup, and partial orderings* $\leqslant$ *on* G *for which* G *is a partially ordered group satisfying*

$$g \leqslant h \text{ in } G,\ c \in Y, \text{ imply } gc \leqslant hc \text{ in } X.$$

The proof of this uses the fact the partial ordering on G comes from P/A, so that $1 \leqslant g^{-1}h$ for $h,g \in G$ if and only if $(1,1) \leqslant (1 \wedge g^{-1}h1, g^{-1}h)$ in P. It follows that $1 = 1 \wedge g^{-1}h1$, so for $c \in Y$,

$$(c,1) = (1,1)(c,1) \leqslant (1,1 \wedge g^{-1}h1)(c,1) = (1 \wedge g^{-1}hc, g^{-1}h),$$

and finally $c \leqslant g^{-1}hc$ in X.

The relevant condition for the corresponding theorem in which L is regular is

$$g \leqslant h \text{ in } G,\ c \in Y, \text{ imply } hc \leqslant gc \text{ in } Y.$$

Combining these dual results, it follows [Blyth 1974a, McFadden 1975] that if R and L are both regular, then the only partial orderings under which P is a strong integrally closed Dubreil-Jacotin inverse semigroup are those for which

$$g \leqslant h \text{ in } G,\ c \in Y, \text{ imply } gc = hc \text{ in } X.$$

EXAMPLE 3. Let P be the bicyclic semigroup $\langle p,q \rangle$ [Clifford and Preston 1961b]. Define the partial ordering $\leqslant$ on P by

$$q^r p^s \leqslant q^u p^v \text{ if and only if } r - s \leqslant u - v \text{ and } v \leqslant s.$$

Then P is a strong integrally closed Dubreil-Jacotin inverse semigroup in which L, but not R, is regular.

Throughout the preceding discussion of regular Dubreil-Jacotin semigroups it was assumed that the maximum element m satisfying the inequality $x^2 \leqslant x$ was the identity element, and it was this assumption that permitted the use of the structure theory of E-unitary inverse semigroups. In general m is not the idenity element, but there is a weaker condition which allows the use of the theory of orthodox semigroups.

An element u of a semigroup S is called a middle unit if $xy = xuy$ for all x,y in S. McAlister and Blyth [1978b] have shown that if S is a regular Dubreil-Jacotin semigroup in which m, as above, also satisfies $x^2 = xmx$ for each x in S, then S is orthodox and m is a middle unit for S.

Let S be a partially ordered regular semigroup. We shall say that the ordering of S is amenable if $x \leqslant y$ in S implies that there exist inverses x', x'' of x and y', y'' of y satisfying

$$xx' \leqslant yy' \quad \text{and} \quad x''x \leqslant y''y.$$

Clearly this is a generalization of the concept of regularity mentioned above for R and L in an inverse semigroup. Now let S be a partially ordered semigroup which is orthodox, and denote by Y the finest inverse semigroup congruence on S. We say that S splits strongly if

(1) S/Y can be ordered in such a way that the canonical mapping $\phi: S \to S/Y$ is isotone,

(2) there is an isotone homomorphism $\theta: S/Y \to S$ such that $\theta\phi$ is the identity map on S/Y.

We then have the following theorem [McAlister and Blyth, 1978b].

THEOREM 7. *Let S be a regular Dubreil-Jacotin semigroup and let m be the maximum element of S satisfying the inequality $x^2 \leqslant x$. If $x^2 = xmx$ for each x in S then S is orthodox. Moreover, S splits strongly, m is a middle unit for S, the band of idempotents of S is normal, and every element of S has a greatest inverse. Furthermore, if S is amenably ordered then for each x in S the meet $xm \wedge mx$ exists and is x.*

To prove this, one recalls that under the hypotheses of the theorem, m is the maximum idempotent of S. Then, as in the proof of theorem 4, mSm is an inverse subsemigroup of S in which the identity element m is the maximum idempotent. It follows that $x^2 = xmx$ for each x in S exactly when

(i) mSm meets each Y-class in a singleton, hence determining a transversal of the Y-classes which is actually an inverse subsemigroup T of S, or

(ii) Y is a closure equivalence, with mxm the maximum element in the Y-class of x.

Under these equivalent conditions S can be shown to split strongly and, apart from the last, the other statements of the theorem follow. In particular, the greatest inverse of $x \in S$ is the inverse in T of mxm, while the band of idempotents of S is normal because if x,y,z,w are idempotents, then

$$xyzw = x.mym.mzm.w = x.mzm.mym.w = xzyw \ .$$

The proof of the last statement uses a thoerem [McAlister and Blyth, 1978b] which describes the structure of a strongly split amenably ordered orthodox semigroup in terms of its band of idempotents and the inverse subsemigroup T.

References

1960 I. Molinaro, Demi-groupes résidutifs. *J. Math. pures et appl.*, série 9, 39 (1960), 319-56.

1961a I. Molinaro, Demi-groupes résidutifs II, ibid, 40 (1961), 43-110.

1961b A.H. Clifford and G.B. Preston, *The algebraic theory of semigroups.* Amer. Math. Soc., Math. Surveys No.7, Vols 1 and 2.

1970 D.B. McAlister and L. O'Carroll, On *B*-nomal semigroups. *J. London Math. Soc.* (2), 2 (1970), 679-88.

1971a T. Saitô, Ordered inverse semigroups. *Trans. Amer. Math. Soc.*, 153 (1971), 99-138.

1971b R. McFadden and L. O'Carroll, *F*-inverse semigroups. *Proc. London Math. Soc.*, 22 (1971), 652-66.

1972a E.A. Behrens, Partially ordered, completely simple semigroups. *Journal of Algebra*, 23 (1972), 413-37.

1972b T.S. Blyth and M.F. Janowitz, *Residuation Theory.* International Series of Monographs in Pure and Applied Mathematics, Vol.102, Pergamon Press, 1972.

1973 R. McFadden, On the structure of certain *A*-nomal semigroups. *J. London Math. Soc.* (2), 6 (1973), 543-52.

1974a T.S. Blyth, Dubreil-Jacotin inverse semigroups. *Proc. Roy. Soc. Edinburgh (A)*, 71 (1974), 345-60.

1974b T.S. Blyth, The structure of certain integrally closed semigroups. *Semigroup Forum*, 8 (1974), 104-24.

1974c D.B. McAlister, Groups, semilattices and inverse semigroups II, *Trans. Amer. Math. Soc.*, 196 (1974), 350-70.

1974d S. Eilenberg, *Automota, Languages and Machines, Vol.A*, Academic Press, 1974.

1975 R. McFadden, Proper Dubreil-Jacotin inverse semigroups. *Glasgow Mathematical Journal*, 16 (1975), 40-51.

1976 J.M. Howie, *An introduction to semigroup theory*, L.M.S. Monographs, London, Academic Press, 1976.

1978a A. Batbedat, γ-demi-groupes, demi-modules, produits demi-directs, *Tagungüber Halbgruppentheorie, Oberwolfach* (1978).

1978b D.B. McAlister and T.S. Blyth, Split orthodox semigroups. *Journal of Algebra* 51 (1978), 491-525.

TO BE PUBLISHED

[A] E.A. Behrens, Semigroups with a non-commutative arithmetic. *Semigroup Forum*. To appear.

[B] D.B. McAlister, Amenably ordered inverse semigroups. *Journal of Algebra*. To appear.

THE FREE ELEMENTARY * ORTHODOX SEMIGROUP

H. E. Scheiblich

Department of Mathematics, Computer Science, and Statistics,
University of South Carolina, Columbia, S.C., U.S.A., 29208

Abstract

A * orthodox semigroup is a unary semigroup $(S,\cdot,*)$ which satisfies the axioms (1) $x^{**} = x$, (2) $xx^*x = x$, (3) $(xy)^* = y^*x^*$, and $xx^*yy^*zz^* \in E(S) = E$. Such a semigroup is orthodox in the usual sense that $EE \subset E$. Since * orthodox semigroups are equationally defined, they form a variety. This paper characterizes the free * orthodox semigroup F on a single generator x.

An idempotent e is a projection provided $e^* = e$. There is a unique projection in each $\mathcal{R}$ class and in each $\mathcal{L}$ class. Each element $w \in F$ is given a canonical form by locating the projection in its $\mathcal{R}$ class and its $\mathcal{L}$ class. The characterization is achieved by describing the multiplication between these canonical forms.

As a corollary, the band $E(F)$ is described. This band is regular in the sense that it satisfies the regularity equation $axaya = axya$.

1. Introduction

Given a semigroup S, two elements a and b are inverses if $aba = a$ and $bab = b$. When a and b are inverses, both ab and ba are idempotents. To identify several classes of semigroups, a semigroup S is *regular* if every element has an inverse, *orthodox* if S is regular and the set E of idempotents forms a subsemigroup, and *inverse* if S is regular and the set E of idempotents is a commutative subsemigroup.

The class of regular semigroups does not contain free objects [1968], and neither does the class of orthodox semigroups [A]. Thus, neither of these classes forms a variety. In the case of inverse semigroups, the condition that E is a commutative subsemigroup forces

ISBN 0-12-319450-4

each element to have a unique inverse [1961]. Thus, it is possible to introduce a unary operation $a \to a^*$, taking a onto the inverse of a. Schein [1963] has found that it is possible to define with equations the class of inverse semigroups by using the usual binary operation and the unary operation $a \to a^*$, thus showing that the class of inverse semigroups forms a variety. His equations are:

(1) $(xy)z = x(yz)$,

(2) $x^{**} = x$,

(3) $(xy)^* = y^*x^*$,

(4) $x = xx^*x$, and

(5') $(xx^*)(x^*x) = (x^*x)(xx^*)$.

In [1978], more structure is added to the classes of regular and orthodox semigroups in order to make them varieties. A semigroup is ** regular* if it satisfies axioms (1) - (4). In a * regular semigroup, x^* is an inverse for x since $x^*xx^* = x^*(x^*)^*x^* = x^*$. In the presence of axioms (1) - (4), the set E of idempotents forms a subsemigroup if and only if S also satisfies the following equation:

(5) $[(xx^*)(yy^*)(zz^*)]^2 = [(xx^*)(yy^*)(zz^*)]$

Such semigroups are called ** orthodox* semigroups. The purpose of this note is to describe the free * orthodox semigroup on a singleton set $\{x\}$.

The free inverse semigroup on a singleton set $\{x\}$ has been described as a certain subsemigroup of $B \times B$ where B is the bicyclic semigroup [1971]. A glance at the proof will show that each element a can be written canonically in the form $a = x^p x^{-n} x^s$ where $x^{-n} = (x^{-1})^n = (x^*)^n$ and $n \geqslant p, s \geqslant 0$. Thus $a = (x^p x^{-p})x^{-(n-p)}x^s$. The initial idempotent $x^p x^{-p}$ identifies the R-class to which a belongs. Similarly, $a = x^p x^{-(n-s)}(x^{\ s} x^s)$ and a belongs to the L-class which contains $x^{-s}x^s$. The characterization to be presented here for * orthodox semigroups is along the lines of finding the R and L-class to which a belongs.

Identifying the structure of a * orthodox semigroup S which is generated by a single element x will be pursued in the following way: (A) find the idempotents of S, (B) find a canonical form for each element of S, and (C) describe the multiplication between these canonical forms.

2. The structure of a * orthodox semigroup generated by *x*

(A) *The idempotents.*

In the case of an inverse semigroup generated by x, it was useful to consider the idempotents $e_n = x^n x^{-n}$ and $f_n = x^{-n}x^n$. It is easily shown that $e_1 \geqslant e_2 \geqslant \dots$ and $f_1 \geqslant f_2 \geqslant \dots$. Since idempotents commute in an inverse semigroup, then $e_r f_s = f_s e_r$. It turned out that every idempotent can be written in the form $e_r f_s$ where $r, s \geqslant 0$.

In the present case of a * orthodox semigroup, $(x^*)^n = (x^n)^*$ by the involution property $(xy)^* = y^*x^*$. Thus, there will be no confusion in abbreviating $(x^*)^n = (x^n)^*$ as x^{-n}. Let $e_n = x^n x^{-n}$ and $f_n = x^{-n}x^n$. Again it is easy to check that $e_1 \geqslant e_2 \geqslant \dots$ and $f_1 \geqslant f_2 \geqslant \dots$. For convenience, put $x^0 = 1 \in S^1$ and $e_0 = f_0 = 1 \in S^1$. For future reference, notice that the operation * leaves e_n fixed; for $e_n^* = (x^n x^{-n})^* = (x^{-n})^*(x^n)^* = x^n x^{-n} = e_n$. Similarly, $f_n^* = f_n$. In an orthodox semigroup, idempotents do not in general commute and so the relationship $e_r f_s = f_s e_r$ will not in general hold. It will turn out, however, that every idempotent can be written as a finite sequence of e_i's and f_j's. Products are not at all unique, however, since some products can be shortened to a product containing fewer terms. If a product has two adjacent e terms, then the product could be shortened since $e_r e_t = e_{\max\{r,t\}}$. A similar statement holds for f terms. Sequences (products) in which e and f terms alternate will be called alternating sequences. The next lemma and its corollary are the main tools for reducing alternating products.

LEMMA 2.1. *When $t \leqslant r$ and $u \leqslant b$, then $e_r f_u e_t f_b = e_r f_b$.*

Proof. $e_r f_u e_t f_b = (e_r e_t) f_u e_t (f_u f_b) = e_r (e_t f_u e_t f_u) f_b = e_r (e_t f_u) f_b = (e_r e_t)(f_u f_b) = e_r f_b$.

COROLLARY 2.2. *Let $r, b > 0$. Let α be a sequence of e_i's and f_j's such that*

(1) *when e_t is a term of α, then $t \leqslant r$, and*

(2) *when f_u is a term of α, then $u \leqslant b$.*

Then $e_r \alpha f_b = e_r f_b$.

Proof. If this were not so, then there would be a sequence α of shortest possible length which satisfies conditions (1) and (2), but $e_r \alpha f_b \neq e_r f_b$. It follows immediately from the minimum length of α that $\alpha = f_{u_1} e_{t_1} f_{u_2} e_{t_2} \ldots f_{u_k} e_{t_k}$ for some integer k with $k \geqslant 2$. Now consider $\beta = e_r f_{u_1} e_{t_1} f_{u_2}$. Since $t_1 \leqslant r$, then $u_2 < u_1$ otherwise β could be shortened to $e_r f_{u_2}$ by lemma 2.1 and hence α could be shortened. Now let $\gamma = f_{u_1} e_{t_1} f_{u_2} e_{t_2}$. Since $u_2 < u_1$, then $t_2 < t_1$ otherwise γ could be shortened to $f_{u_1} e_{t_2}$ by the e-f dual to lemma 2.1 and hence α could be shortened. Continuing this, it follows that $b < u_k$. This contradicts condition (2).

On the other hand, consider the sequence $e_r f_s e_t f_u$ where $r < t$ and $s < u$. A little experimentation will convince the reader that it will not be possible to reduce the length of this product of length 4. This motivates the next definition.

Let U be the set of all alternating sequences w of e_i's and f_j's such that

$$\begin{aligned} &(1) \quad \text{when } e_r f_s e_t \text{ is a subsequence of } w, \text{ then } r < t, \\ &(2) \quad \text{when } f_s e_t f_u \text{ is a subsequence of } w, \text{ then } s < u. \end{aligned} \tag{2.3}$$

The set U contains the sequences of length 1, e_n and f_n. Also for convenience U contains the empty sequence $\square$. Some elements of U are: e_5, $e_2 f_4 e_4 f_5 e_5 f_9$, $e_2 f_1 e_6$. Now let $V = U^* = \{w^*: w \in U\}$. For example, since $e_2 f_1 e_6 \in U$, then $(e_2 f_1 e_6)^* = e_6 f_1 e_2 \in V$. It will not be possible to reduce the products in U or V to products of shorter length.

Now let $w \in U$, say w has length k. Then the notation $w\, e_r f_s = w + e_r f_s \in U$ will mean that w ends with an f term and that the juxtaposition product $w\, e_r f_s$ is an element of U of length $k + 2$. Similar analogous statements apply to $w\, f_s e_r = w + f_s e_r \in U$, $w\, e_n = w + e_n \in U$, $e_r f_s\, v = e_r f_s + v \in V$ when $v \in V$, and so on. This is in opposition to the following situation: $w = e_2 f_1 e_6$, $e_r f_s = e_7 f_3$. Here $w\, e_r f_s = (e_2 f_1 e_6)(e_7 f_3) = e_2 f_1 e_6 e_7 f_3 = e_2 f_1 e_7 f_3 \in U$, but

$w\ e_rf_s \neq w + e_rf_s$.

Suppose now that $w \in U$, $v \in V$, $w\ e_rf_s = w + e_rf_s \in U$, and $e_rf_s\ v = e_rf_s + v \in V$. The sequence $w\ e_rf_s\ v$ contains no subsequence of the type in lemma 2.1. Thus, it will not be possible to reduce this idempotent to a product of shorter length. A similar statement applies to $w\ e_rf_se_r\ v$ where $w\ e_rf_s = w + e_rf_s \in U$ and $r_se_r\ v = f_se_r + v \in V$. It will emerge that every idempotent, besides those in U and V, is of the form (1) $w\ e_rf_s\ v$ where $w \in U$, $v \in V$, and $w\ e_rf_s = w + e_rf_s$, $e_rf_s\ v = e_rf_s + v$, or (2) $w\ e_rf_se_r\ v$ where $w\ e_rf_s = w + e_rf_s$ and $f_se_r\ v = f_se_r + v$, or (3) the duals to (1) and (2).

In a * orthodox semigroup S, an idempotent e is called a *projection* provided $e = e^*$. The projections play a vital rose in the structure of S, since there is a unique projection in each R-class and in each L-class.

As noted previously, $e_n^* = e_n$ and $f_n^* = f_n$ and so e_n and f_n are projections. Thus, $e_rf_se_r$ is also a projection. Further, $w(e_rf_se_r)w^*$ is a projection when $w \in U$ and $w\ e_rf_s = w + e_rf_s \in U$. It will emerge that a complete list of all projections is (1) e_n, (2) $e_rf_se_r$, (3) $w\ e_rf_se_r\ w^*$ as above, and (4) the e-f duals of (1), (2), and (3).

The projections e_n and f_n are D-related since $(e_n, x^n) \in R$ and $(x^n, f_n) \in L$. When $r, s > 0$ and $r + s = n$, then $(e_n, e_rf_se_r) \in D$. To see this, let $a = e_nx^se_r$. Then $aa^* = e_nx^se_re_rx^{-s}e_n = e_nx^sx^rx^{-r}x^{-s}\ e_n$ $= e_ne_{r+s}e_n = e_n$. Thus, $(e_n, a) \in R$. Notice that $a = x^nx^{-n}x^se_r =$ $= x^{s+r}x^{-(r+s)}x^se_r = x^se_rf_se_r$. Thus $a^*a = e_rf_se_rx^{-s}x^se_rf_se_r =$ $= (e_rf_s)^3e_r = e_rf_se_r$. Thus, $(a, e_rf_se_r) \in L$. Presently there will be enough lemmas available to show that $(e_rf_se_r,\ w(e_rf_se_r)w^*) \in D$ whenever $w \in U$ and $w\ e_rf_s = w + e_rf_s \in U$. Thus, the D-classes will be the sets D_1, D_2, ... where $D_n = D_{e_n}$.

In the light of the previous paragraph, it will be useful to partition the sets U and V. When $r, s > 0$ and $r + s > 0$, let

$$U(r,s) = \{w \in U\colon\ w \text{ terminates with } e_rf_s \text{ or } f_se_r\}$$
$$U_n = \bigcup_{r+s=n} U(r,s), \qquad V(r,s) = U(r,s)^*, \qquad V_n = U_n^*. \tag{2.4}$$

When $n > 0$, $U(n,0) = \{e_n\}$ and $U(0,n) = \{f_n\}$.

It is going to be necessary to multiply products like BC where $B \in V(r,s)$ and $C \in U(a,b)$. Using the involution property it will be enough to consider the case $r + s \leqslant a + b$. An example would be a product of the form $(e_r f_s\ v)(w\ f_b e_a)$. The next two lemmas are to handle this situation.

Let $C \in U$. Suppose that $C \in U(a,b)$ where $r < a$ and $s < b$. Define $C(r,s)$ to be the longest final segment of C such that

$$\begin{aligned}&(1)\quad \text{whenever } e_t \text{ is a term of } C(r,s), \text{ then } r < t, \text{ and}\\ &(2)\quad \text{whenever } f_u \text{ is a term of } C(r,s), \text{ then } s < u.\end{aligned} \tag{2.5}$$

LEMMA 2.6. $(e_r f_s\ v)C = e_r f_s[\,C(r,s)]$.

Proof. Let t,u be minimal such that e_t, f_u are terms of C and $r < t$, $s < u$. There are two cases depending on which of e_t and f_u appears first.

(1) Assume that e_t appears first. Thus C is of the form $\lambda\ f_\beta e_t\ \tau\ e_\alpha r_u\ \delta$ where some or all of λ, τ, δ may be empty. Then $C(r,s) = e_\alpha f_u\ \delta$ and $e_r f_s C(r,s) = e_r f_s e_\alpha f_u\ \delta$. Also $(e_r f_s\ v)C =$ $= e_r(f_s\ v\ \lambda\ f_\beta e_t\ \tau\ e_\alpha)f_u\ \delta = e_r(f_s e_\alpha)f_u\ \delta$ by the dual to corollary 2.2.

(2) Assume now that $C = \lambda\ e_\alpha f_u\ \tau\ f_\beta e_t\ \delta$. Then $C(r,s) = f_\beta e_t\ \delta$ and so $e_r f_s C(r,s) = e_r f_s f_\beta e_t\ \delta = e_r f_\beta e_t\ \delta$. Also $(e_r f_s\ v)C =$ $= (e_r f_s\ v\ \lambda\ e_\alpha f_u\ \tau\ f_\beta)e_t\ \delta = (e_r f_\beta)e_t\ \delta$ by corollary 2.2.

Consider again the product BC where $B \in V(r,s)$, $C \in U(a,b)$, and $r + s \leqslant a + b$. The case $r < a$ and $s < b$ is managed by lemma 2.6. In the opposite case either $a \leqslant r$ or $b \leqslant s$. The head h of B is either $e_r f_s$ or $f_s e_r$ and the tail t of C is either $e_a f_b$ or $f_b e_a$. Thus, $B = hv$ where $v \in V$ and $hv = h + v \in V$ and $C = wt$ where $w \in U$ and $wt = w + t \in U$.

LEMMA 2.7. $hvwt = ht$.

Proof. Since $a \leqslant r$ or $s \leqslant b$, assume without loss of generality that $a \leqslant r$. Since $r + s \leqslant a + b$, then $s \leqslant b$. There will be four cases depending on the forms of h and t. Corollary 2.2 and its dual will be used freely.

(1) $[h = e_r f_s,\ t = e_a f_b]\ hvwt = e_r(f_s\, v\, w\, e_a) f_b = e_r f_b = e_r(f_s e_a) f_b = ht$.

(2) $[h = e_r f_s,\ t = f_b e_a]\ hvwt = [e_r(f_s\, v\, w) f_b]\, e_a = [e_r f_b]\, e_a = e_r f_s f_b e_a = ht$.

(3) $[h = f_s e_r,\ t = e_a f_b]\ hvwt = f_s [e_r(v\ w\ e_a) f_b] = f_s [e_r f_b] = f_s e_r e_a f_b = ht$.

(4) $[h = f_s e_r,\ t = f_b e_a]\ hvwt = f_s [e_r(v\ w) f_b]\, e_a = f_s [e_r f_b]\, e_a = ht$.

Lemma 2.7 will now be used to check the statements made earlier that $(e_r f_s e_r,\ w\ e_r f_s e_r\ w^*) \in \mathcal{D}$. Let $a = e_r f_s e_r\ w^*$. Then $aa^* = e_r f_s e_r\ w^* w\ e_r f_s e_r = e_r\ [f_s e_r\ w^* w\ e_r f_s\ ] e_r = e_r [(f_s e_r)(e_r f_s)] e_r = (e_r f_s e_r)^2 = e_r f_s e_r$. It follows immediately that $a^* a = w\ e_r f_s e_r\ w^*$. Thus, $e_r f_s e_r\ \mathcal{R}\ a\ \mathcal{L}\ w\ e_r f_s e_r\ w^*$.

(B) *Canonical Forms.*

As in the inverse semigroup case, Green's relation $\mathcal{H}$ will be the identity. Thus, when $a \in S$, then $a = \{a\} = H_a = R_a \cap L_a$. Since aa^* is the unique projection in R_a and $a^* a$ is the unique projection in L_a, it is possible to represent a by $[aa^* : a^* a] \in R_a \times L_a$. More generally, any element a can be represented uniquely in the form $[g:h]$ where g and h are projections in the same $\mathcal{D}$-class.

Recall that the projections have been identified as elements of the form e_n, $e_r f_s e_r$, $w\ e_r f_s e_r\ w^*$, and their duals. Suppose then, for example, that $[g:h] = [u\ e_p f_q e_p\ u^* :\ w\ f_s e_r f_s\ w^*]$ where $p + q = r + s$. In the element a that $[g:h]$ represents, the projection g is to carry the code for the $\mathcal{R}$-class $R_a = R_g$. Thus, only the initial idempotent segment $u\ e_p f_q$ of g will be needed. Similarly, only the final idempotent segment $e_r f_s\ w^*$ of h will be needed. Taking advantage of this, $[g:h]$ may be abbreviated as $[u\ e_p f_q :\ e_r f_s\ w^*]$. The advantage to this abbreviation is that $[g:h] \equiv [u\ e_p f_q :\ e_r f_s\ w^*] \in U(p,q) \times V(r,s)$. Thus, if D_n is the $\mathcal{D}$-class containing e_n and P_n is the set of all projections in D_n, then $D_n = P_n \times P_n \equiv U_n \times V_n$.

It does remain to say, in a canonical way, exactly which element of S is represented by $[A:B] = [g:h] \equiv [u\ e_p f_q :\ e_r f_s\ w^*]$. This will be done in the following table. Here, $p + q = r + s = n$.

$$[e_n : f_n] = (x^n)(x^{-n})(x^n)$$

$$[e_n : e_r f_s \ v] = (x^n)(x^{-r} f_s \ v)$$

$$[e_n : f_s e_r \ v] = (x^n)x^{-n}(x^s e_r \ v) \qquad (2.6)$$

$$[u \ e_p f_q : e_r f_s \ v] = (u \ e_p \ x^{-q})x^n(x^{-r} f_s \ v)$$

$$[u \ e_p f_q : f_s e_r \ v] = (u \ e_p \ x^{-q})(x^s \ e_r \ v)$$

All other cases are dual to these. For example, $[u \ f_q e_p : e_r f_s \ v] = u \ f_q \ x^p x^{-r} f_s \ v$.

Notice that in the element $[u \ e_p f_q : e_r f_s \ v]$, the idempotent $u \, e_p f_q$ is an initial setment and $e_r f_s \ v$ is a final segment. To check this, $[u \ e_p f_q : e_r f_s \ v] = (u \ e_p x^{-q})x^n(x^{-r} f_s \ v) = (u \ e_p \ x^{-q})x^q x^p(x^{-r} f_s \ v) = (u \ e_p f_q)x^p x^{-r} f_s \ v$, and similarly, $[u \ e_p f_q : e_r f_s \ v] = u \ e_p x^{-q} x^s (e_r f_s \ v)$. This also happens in all the other canonical forms. For example, $[u \ e_p f_q : f_s e_r \ v] = (u \ e_p x^{-q})(x^s e_r \ v) = (u \ e_p \ x^{-q})x^s x^r x^{-r} \ v = u \ e_p \ x^{-q} x^{r+s=q+p} x^{-r} \ v = (u \ e_p f_q)x^p x^{-r} \ v$. Similarly, $[u \ e_p f_q : f_s e_r \ v] = u \ e_p \ x^{-r}(f_s e_r \ v)$.

Once again, let $[A, B] = [u \ e_p f_q : e_r f_s \ v] \in U_n \times V_n$. Then $[A:B]*$ $= [(u \ e_p \ x^{-q})x^n(x^{-r} f_s \ v)]* = [(v* \ f_s \ x^r)x^{-n}(x^q e_p \ u*)] = [v* \ f_s e_r : f_q e_p \ u*] = [B*:A*]$. A check will reveal that this happens in general: for all $[A:B] \in D_n$, $[A:B]* = [B*:A*]$.

(C) *Multiplication of Canonical Forms.*

The purpose of this section is to present a theorem which tells how to multiply canonical forms. Before proceeding to that, one more operator will be needed.

It is known that when S is an orthodox semigroup with band of idempotents E, $x \in S$, x' is an inverse of x, and $e \in E$, then $x \ e \ x' \in E$. Thus, when S is * orthodox $x \ E \ x* \subset E$. The next lemmas are to handle the operation $e \to x \ e \ x*$.

LEMMA 2.7. *(a) For* $r \geq 0$, $x \ e_r \ x^{-1} = e_{r+1}$,

(b) for $s \geq 1$, $x \ f_s \ x^{-1} = e_1 \ f_{s-1} \ e_1$,

(c) *for* $r,t \geqslant 0$ *and* $s \geqslant 1$, $x(e_r f_s e_t)x^{-1} =$
$= e_{r+1}f_{s-1}e_{t+1}$,

(d) *for* $k \geqslant 0$; $r,t \geqslant 0$; *and* $s \geqslant k$; $x^k(e_r f_s e_t)x^{-k} =$
$= e_{r+k}f_{s-k}e_{t+k}$.

Proof. Part (a) follows immediately from the definition of $e_r = x^r x^{-r}$. Part (b) is almost as easy using $f_s = x^{-s}x^s$. For (c), $x(e_r f_s e_t)x^{-1} = x[e_r(f_1 f_s f_1)e_t]x^{-1} = x[e_r(x^{-1}x\ f_s\ x^{-1}x)e_t]x^{-1} =$ $= (x\ e_r\ x^{-1})(x\ f_s\ x^{-1})(x\ e_t\ x^{-1}) = e_{r+1}f_{s-1}e_{t+1}$. Finally, part (d) follows from (c) by induction.

It is not hard now to compute $x^k u x^{-k}$ for any alternating sequence u for which each f subscript is at least k. Simply (1) begin and end the sequence with an e term by adding e_0 if necessary, and (2) add k to each e subscript and add $-k$ to each f subscript. For example, $x^3(e_2f_4e_3f_6)x^{-3} = x^3(e_2f_4e_3f_6e_0)x^{-3} = e_5f_1e_6f_3e_3$.

The dual to lemma 2.7(d) will also be useful. This dual says that $x^{-k}(f_s e_r f_u)x^k = f_{s+k}e_{r-k}f_{u+k}$. Notice that this operation $u \to x^{-k}\ u\ x^k$ adds $-k$ to the subscript of e terms and k $(= -(-k))$ to the subscript of f terms.

Let k be an integer (positive, negative, or zero). Let α be an alternating sequence such that each e subscript and each f subscript is at least $|k|$. Then the operator E^k operates on α by

$E^k(\alpha)$ is the alternating sequence obtained from α by adding k to each e subscript and adding $-k$ to each f subscript. (2.8)

For, example, $E^5(f_7e_1f_6e_4) = f_2e_6f_1e_9$ and

$$E^{-4}(e_5f_6e_9) = e_1f_{10}e_5.$$

COROLLARY 2.9. *(1) If an alternating sequence* α *begins and ends with an* e *term,* $k > 0$, *and each* f *subscript is greater than* k, *then* $x^k\ \alpha\ x^{-k} = E^k(\alpha)$.

(2) If an alternating sequence α *begins and ends with an* f *term,* $k > 0$, *and each* e *subscript is greater than* k, *then* $x^{-k}\ \alpha\ x^k = E^{-k}(\alpha)$.

The multiplication between canonical forms will now be described. Notice that it is sufficient, by the involution property, to find $a \cdot c$ where $a \in D_n$, $c \in D_m$ and $n \leqslant m$. Some multiplication of sequences in U is involved. In each case, the multiplication is either juxtaposition or juxtaposition with the collapse of a single term. A similar statement applies to sequences in V.

THEOREM 2.10. *Let* $a = [A:B] \in U(p,q) \times V(r,s) \subset D_n$ *and let* $c = [C:D] \in U(a,b) \times V(c,d) \subset D_m$ *with* $n \leqslant m$. *Then* $a \cdot c =$

(1)	$[r < a,\ s < b]$	$[A\ E^{s-q}(C(r,s)):D]$	$\in D_{a+b}$
(2)	$[r < a,\ s = b]$	$[A\ e_{a+s-q}:D]$	$\in D_{a+b}$
(3)	$[r < a,\ b < s]$	$[A\ e_{a+s-q}:f_{a+s-c}\ D]$	$\in D_{a+s}$
(4)	$[r = a,\ b = s]$	$[A:D]$	$\in D_{a+b}$
(5)	$[r = a,\ s < b]$	$[A\ f_{r+b-p}:D]$	$\in D_{a+b}$
(6)	$[a < r,\ s < b]$	$[A\ f_{r+b-p}:e_{r+b-d}\ D]$	$\in D_{r+b}$.

Proof. First, let us verify that the sequence multiplication is of the type described. Consider cases (2) and (3) where $r < a$. Since $r < a$, then $r + s - q < a + s - q$, i.e., $p < a + s - q$. A is either of the form $u\ e_p f_q$ or of the form $u\ f_q e_p$. In the first form, $A\ e_{a+s-q} = (u\ e_p f_q) e_{a+s-q} = u\ e_p (f_q e_{a+s-q}) \in U(a+s-q,\ q) \subset U_{a+s}$. In the second form, $A\ e_{a+s-q} = (u\ f_q e_p) e_{a+s-q} = (u\ f_q)(e_p\ e_{a+s-q}) = u\ f_q\ e_{a+s-q} \in U(a+s-q,\ q) \subset U_{a+s}$. Similar computations apply to the sequence multiplications in cases (3), (5), and (6). In case (1), suppose $C(r,s)$ has initial e term e_t and initial f term f_u. Thus, $E^{s-q}\ C(r,s)$ has initial e term e_{t+s-q} and initial f term f_{u+q-s}. But $r < t$ and so $r + s - q < t + s - q$, i.e., $p < t + s - q$. Similarly, $s < u$ and so $s + q - s < u + q - s$, i.e., $q < u + q - s$. So, the multiplication $A \cdot E^{s-q}(C(r,s))$ is either juxtaposition or juxtaposition with one term collapsing.

Since there are several choices for each of A, B, C, and D, the proof involves considerable case checking. Here only one case will be checked. Let $a = [A:B] = [u\ e_p f_q : e_r f_s\ v] \in U(p,q) \times V(r,s) \subset D_n$ and let $c = [C:D] = [w\ f_b e_a : e_c f_d\ y] \in U(a,b) \in V(c,d) \subset D_m$ where $n \leqslant m$.

Then $a = (u\ e_p\ x^{-q})x^n(x^{-r}f_s\ v) = u\ e_p x^{-q}x^{s+r}x^{-r}f_s\ v =$
$= u\ e_p x^{-q}x^s(e_r f_s\ v)$ and $c = (w\ f_b x^a)(x^{-c}f_d\ y) = w\ f_b x^a x^{-c}x^{-d}x^d y =$
$w\ f_b\ x^a x^{-(a+b)}x^d y = w\ f_b e_a x^{-b}x^d y.$ Thus
$a\cdot c = u\ e_p x^{-q}x^s(e_r f_s v\ w\ f_b e_a)x^{-b}x^d y.$

(1) $[r < a,\ s < b]$. First, lemma 2.6 says that $(e_r f_s\ v\ w\ f_b e_a) = e_r f_s[(w\ f_b e_a)(r,s)]$. Thus, $ac =$

$$u\ e_p x^{-q}x^s[e_r f_s[(w\ f_b e_a)(r,s)]]x^{-b}x^d y =$$
$$u\ e_p x^{-q}x^s[e_r f_s(w\ f_b e_a)(r,s)]x^{-s}x^{-(b-s)}x^d y =$$
$$u\ e_p x^{-q}[e_{r+s}E^s(w(r,s)f_b e_a)]x^{-(b-s)}x^d y =$$
$$u\ e_p x^{-q}[e_{r+s}\ E^s(w(r,s)f_b)]e_{a+s}x^{-(b-s)}x^d y =$$
$$u\ e_p x^{-q}[E^s(e_r\ w(r,s)f_b)]x^{a+s}x^{-(a+s)}x^{-(b-s)}x^d y =$$
$$u\ e_p x^{-q}[E^s(e_r\ w(r,s)f_b)]x^q x^{a+s-q}x^{-(a+b)}x^d y =$$
$$u\ e_p f_q[E^{s-q}(e_r\ w(r,s)f_b)]x^{a+s-q}x^{-(c+d)}x^d y =$$
$$u\ e_p f_q[E^{s-q}(e_r\ w(r,s)]f_{b+q-s}x^{a+s-q}x^{-c}f_d y =$$
$$[u\ e_p f_q[E^{s-q}(e_r\ w(r,s)]f_{b+q-s}e_{a+s-q} : e_c f_d y] =$$
$$[u\ e_p f_q e_p\ E^{s-q}(w(r,s)f_b e_a) : D]$$
$$[u\ e_p f_q\ E^{s-q}(w\ f_b e_a(r,s) : D]$$
$$[A\ E^{s-q}(C(r,s)) : D].$$

In cases (2) to (6), $(e_r f_s\ v\,w\ f_b e_a) = e_r f_s f_b e_a$ by lemma 2.7.

In cases (2) to (4), $b \leqslant s$ from which $e_r f_s f_b e_a = e_r f_s e_a$. Thus, $a\cdot c =$

$$u\ e_p x^{-q}x^s(e_r f_s e_a)x^{-b}x^d y =$$
$$u\ e_p x^{-q}x^s x^r x^{-r}x^{-s}x^s x^a x^{-a}x^{-b}x^d y =$$
$$u\ e_p x^{-q}\ e_{r+s}x^{a+s}x^{-(a+b)}x^d y =$$
$$u\ e_p x^{-q}x^{a+s}x^{-(c+d)}x^d y \quad (\text{since } r + s \leqslant a + s) =$$

$$u\ e_p x^{-q} x^{a+s} x^{-c} f_d y =$$

(2) $[r < a,\ s = b]$ $u\ e_p x^{-q} x^{q} x^{a+s-q} x^{-c} f_d y =$

$u\ e_p f_q e_{a\ s-q}\ e_c f_d y$

$[A\ e_{a+s-q}:D]$.

(3) $[r < a,\ b < s]$ $u\ e_p x^{-q} x^{a+s} x^{-(a+s)} x^{a+s} x^{-c} f_d y =$

$u\ e_p x^{-q} x^{q} x^{a+s-q} x^{-(a+s)} x^{a+s-c} x^{c} x^{-c} f_d y =$

$u\ e_p f_q e_{a+s-q} : f_{a+s-c}\ e_c f_d y =$

$[A\ e_{a+s-q} : f_{a+s-c}\ D]$.

(4) $[r = a,\ s = b]$ $u\ e_p x^{-q} x^{r+s=a+b} x^{-c} f_d y =$

$u\ e_p f_q : e_c f_d y =$

$[A:D]$.

In cases (5) and (6), $s \leqslant b$. So, $e_r f_s f_b e_a = e_r f_b e_a$. Thus, $ac = u\ e_p x^{-q} x^{s}\ e_r f_b e_a\ x^{-b} x^{d} y =$

$u\ x^{p} x^{-p} x^{-q} x^{s} x^{r} x^{-r} x^{-b} x^{b} x^{a} x^{-a} x^{-b} x^{d} y =$

$u\ x^{p} f_{r+s} x^{-(r+b)}\ e_{a+b}\ x^{d} y =$

$u\ x^{p} x^{-(r+b)} x^{d} y$ (since $r + s,\ a + b \leqslant r + b$) $=$

$u\ x^{p} x^{-(r+b)} x^{r+b} x^{-(r+b)} x^{d} y =$

$u\ x^{p} x^{-p} x^{-(r+b-p)} c^{r+b} x^{-(r+b-d)} x^{-d} x^{d} y =$

$u\ e_p\ x^{-(r+b-p)} x^{r+b} x^{-(r+b-d)} f_d y =$

(5) $[r = a,\ s < b]$ $u\ e_p\ x^{-(r+b-p)} x^{a+b} x^{-(a+b-d)} f_d y =$

$u\ e_p\ x^{-(r+b-p)} x^{a+b} x^{-c} f_d y =$

$[u\ e_p f_{r+b-p} : e_c f_d y] =$

$[u\ e_p f_g f_{r+b-p} : D] =$

$[A\ f_{r+b-p} : D]$.

(6) $[a < r,\ s < b]$ $[u\ e_p f_{r+b-p} : e_{r+b-d} f_d y] =$

$$[u\ e_p f_q f_{r+b-p} : e_{r+b-d}\ e_c f_d y] =$$

$$[A\ f_{r+b-p} : e_{r+b-d}\ D].$$

3. The free elementary * orthodox semigroup

Let $e_1, e_2, \ldots$ and $f_1, f_2, \ldots$ be distinct symbols. Let A be the set of all finite alternating sequences of e_i's and f_j's. Let * be a unary operation on A by $\alpha^* = (\alpha_1\alpha_2\ldots\alpha_n)^* = \alpha_n\alpha_{n-1}\ldots\alpha_1$. Let $U \subset A$ be the set of all those sequences which satisfy condition 2.3 . Let $V = U^* = \{w^*: w \in U\}$.

Define now a partial binary operation on U as follows. Let $A, C \in U$ such that (1) the subscript of the last e term of A is less than the subscript of the first e term of C and (2) the same statement for the subscripts of f terms. If A ends with an e term and C begins with an e term, say $A = u\ e_r$ and $C = e_t\ w$, then the product AC is defined by $AC = (u\ e_r)(e_t\ w) = u\ e_t\ w$. If A ends with an e term and C begins with an f term, then the product AC is simple juxtaposition. The e-f dual statement also applies. In a similar way, a partial binary operation is defined on V.

Now let k be any integer. Define the partial operator E^k on U by [2.8].

Define the sets $U(r,s)$, $V(r,s)$, U_n, V_n by [2.4].

When $C \in U(a,b)$ and $r < a$, $s < b$, define $C(r,s)$ by [2.5].

Let $D_n = U_n \times V_n$ and let $S = \bigcup_{n=1}^{\infty} D_n$.

Define a unary operation * on S by $[A:B]^* = [B^*:A^*]$.

To define a binary operation on S, let $a = [A:B] \in U(p,q) \times V(r,s) \subset D_n$ and let $c = [C:D] \in U(a,b) \times V(c,d) \subset D_m$. If $n \leqslant m$, define $a \cdot c$ by the schedule in theorem 2.10. If $n < m$, let $a \cdot c = (c^* \cdot a^*)^*$ whre $c^* \cdot a^*$ is defined by theorem 2.10.

Finally, let $\{z\}$ be a singleton set and let $\varepsilon: \{z\} \to S$ by $\varepsilon(z) = [e_1 : f_1]$.

THEOREM 3.1. *$(S,\cdot,*,\varepsilon)$ is a free * orthodox semigroup on $\{z\}$.*

Outline of Proof. It must be shown that $\cdot$ is an associative binary operation on S. That $\cdot$ is a binary operation is the argument which begins the proof of theorem 2.10. That $\cdot$ is associative involves considerable case checking and will not be included here.

It must be checked that the unary operation * satisfies axioms (2) - (5) of the introduction. It is a routine check that (2) - (4) are satisfied. For (5), first show that $[A:B] \in U(p,q) \times V(r,s)$ is an idempotent if and only if $p = r$ and $q = s$. Then show that the product of two idempotents is idempotent.

At this stage, $(S,\cdot,*)$ is a * orthodox semigroup.

To prove that S is free, let T be any * orthodox semigroup and let $f:\{z\} \to T$, say $f(z) = x$. Then $\theta: S \to T$ is defined by $\theta([A:B)]$ being as in table [2.6]. Note that $(\theta \circ \varepsilon)(z) = \theta[e_1:f_1] = x^1x^{-1}x^1 = x =$ $= f(z)$. That θ is a $\cdot$ homomorphism is the proof of theorem 2.10. That θ preserves * is a routine check. Finally, θ is the only homomorphism from S to T such that $\theta \circ \varepsilon = f$ since $[e_1:f_1]$ generates S as a * semigroup.

The characterization is based on R and L classes. Thus, it is easy to give Green's relations. Let $a = [A:B] \in U(p,q) \times V(r,s)$ and let $c = [C:D] \in U(a,b) \times V(c,d)$ where $p + q = r + s = n$, $a + b =$ $= c + d = m$.

COROLLARY 3.2. *(1) $(a,c) \in R$ if and only if $A = C$ and $(a,c) \in L$ if and only if $B = D$.*

(2) $H = i$, the identity relation.

(3) $(a,c) \in D$ if and only if $n = m$.

The band E of idempotents has its maximal rectangular subbands to the sets $U(r,s) \times V(r,s)$, hereafter denoted $R(r,s)$. Thus, the greatest semilattice image of E is $(N \times N) - (0,0)$ where $N = \{0,1,2,...\}$. The semilattice $(N \times N) - (0,0)$ is also the semilattice of the free inverse semigroup on a singleton set.

The projections in the rectangular band $R(r,s)$ are the elements $[w\ e_rf_s:f_se_r\ w^*] \equiv w\ e_rx^{-s}x^se_r\ w^* = w(e_rf_se_r)w^*$ and the dual

$[w\ f_s e_r : e_r f_s\ w^*] = w(f_s e_r f_s)w^*$. An idempotent has either the form (1) $[w\ e_r f_s : e_r f_s\ v] \equiv w\ e_r x^{-s} x^{s+r} x^{-r} f_s v = w(e_r f_s e_r f_s)v = w(e_r f_s)v$, or (2) $[w\ e_r f_s : f_s e_r v] \equiv w\ e_r x^{-s} x^{s}\ e_r\ v = w(e_r f_s e_r)v$, or (3) the duals to (1) and (2).

When restricted to the idempotents, the multiplication of theorem 2.10 becomes more simple. To give the multiplication in any band, it is sufficient to give the structure maps $\phi_{\alpha,\beta}^{(i,\mu)}$ and $\psi_{\beta,\alpha}^{(i,\mu)}$ where $\alpha \geqslant \beta$ and $(i,\mu) \in R_\alpha$ [1977]. In the case of a * regular band it is sufficient, because of the involution property, to give the left structure maps $\phi_{\alpha,\beta}^{(i,\mu)}$. Let $(r,s) = \alpha \geqslant \beta = (a,b)$ so that $r \leqslant a$, $s \leqslant b$ and let $[A{:}B] \in R(r,s)$. The relevant part of theorem 2.10 will be stated in its simplified form. Let $[C{:}D] \in R(a,b)$. Thus, $[A{:}B][C{:}D] = [[\phi_{\alpha,\beta}^{[A,B]}(C)]:D]$. Note that $(p,q) = (r,s)$ and $(a,b) = (c,d)$.

(1)	$[r < a,\ s < b]$	$[AE^{s-q}C(r,s):D]$	$= [AC(r,s):D]$	$\in R(a,b)$
(2)	$[r < a,\ s = b]$	$[A\ e_{a+s-q}:D]$	$= [A\ e_a:D]$	$\in R(a,b)$
(4)	$[r = a,\ s = b]$	$[A:D]$	$= [A:D]$	$\in R(a,b)$
(5)	$[r = a,\ s < b]$	$[A:f_{r+b-p}:D]$	$= [A\ f_b:D]$	$\in R(a,b)$

Petrich shows that a band B is regular (satisfies the equation $a\,x\,a\,y\,a = a\,x\,y\,a$) if and only if the left structure map $\phi_{\alpha,\beta}^{(i,\mu)}$ is independent of μ and the right structure map $\psi_{\beta,\alpha}^{(i,\mu)}$ is independent of i. Thus, in the case of a * regular band B, B is regular if and only if $\phi_{\alpha,\beta}^{(i,\mu)}$ is independent of μ. In the notation of the previous paragraph, this translates to $[A{:}B][C{:}D]$ is independent of B. A glance at the table shows that the product is independent of B and so the band E is regular.

Some of this will now be stated as a corollary to theorem 3.1.

COROLLARY 3.3. *Let S be a free * orthodox semigroup on a singleton set $\{z\}$. Then the band E of S has greatest semilattice image $(N \times N) - (0,0)$. The band E is regular.*

References

1961 A.H. Clifford and G.B. Preston, *The Algebraic Theory of Semigroups*, Math. Surveys No.7, Amer. Math. Soc., Providence, Vol.1, 1961.

1963 B.M. Schein, On the theory of generalized groups (Russian). *Dokl. Akad. Nauk SSR* 153 (1963), 296-299.

1968 D.B. McAlister, A homomorphism theorem for semigroups. *J. London Math. Soc.*, 43 (1968), 355-366.

1971 H.E. Scheiblich, A characterization of a free elementary inverse semigroup. *Semigroup Forum*, Vol.2 (1971), 76-79.

1977 M. Petrich, *Lectures in Semigroups*, Akademie-Verlag Berlin, 1977.

1978 T.E. Nordahl and H.E. Scheiblich, Regular * semigroups. *Semigroup Forum*, Vol.16 (1978), 369-377.

TO BE PUBLISHED

[A] P.G. Trotter, Free objects in some categories of semigroups. To appear.

ON THE REGULARITY OF CERTAIN SEMIGROUP ALGEBRAS

W. D. Munn

Department of Mathematics, University of Glasgow, University Gardens, Glasgow, G12 8QW, U.K.

Abstract

For two classes of regular 0-simple semigroups, each containing the class of completely 0-simple semigroups, it is shown that if the contracted algebra over a field F of a member of the class is regular (in the sense of von Neumann) then S has to be completely 0-simple. In addition, necessary and sufficient conditions are given for the regularity of the contracted algebra over F of a completely 0-simple semigroup.

The classical theorem of Maschke provides a necessary and sufficient condition for the algebra of a finite group G over a field F to be semisimple. A generalisation of this theorem giving necessary and sufficient conditions for the semisimplicity of the algebra of a finite semigroup S over F was obtained, independently, by Munn [1955] and Ponizovskiĭ [1956] (see Chapter 5 of Clifford and Preston [1961]). Since a finite-dimensional algebra is regular if and only if it is semisimple these results yield necessary and sufficient conditions for the regularity of finite-dimensional group algebras and semigroup algebras respectively.

If G and S are no longer assumed to be finite the situation is more complicated. The problem of finding necessary and sufficient conditions for a group algebra to be regular attracted the attention of several authors before a complete solution (quoted as lemma 1.2 below) was obtained by Villamayor [1959] and Connell [1963]. A study of the analogous (and apparently harder) problem for semigroup algebras was begun by Weissglass [1970c]. He proved, in particular, that if a semigroup algebra is regular then the semigroup itself must be regular and all its subgroups locally finite.

Now suppose that the algebra of a semigroup S over a field F is regular. Let Q denote any principal factor of S. Then, as is easily

ISBN 0-12-319450-4

seen, the contracted algebra of Q^0 over F is regular: moreover, since S is regular, Q^0 is a regular 0-simple semigroup. For this reason it is natural to seek necessary and sufficient conditions for the regularity of the contracted algebra of a regular 0-simple semigroup. It should be noted, however, that a knowledge of such conditions would not suffice to provide a solution of the general problem: any free inverse semigroup is an example of a semigroup S whose algebra over F is not regular although the contracted algebra of Q^0 over F is regular for each principal factor Q of S.

This paper is concerned solely with regular 0-simple semigroups. In §2, two extensive classes of regular 0-simple semigroups are considered, each containing the class of completely 0-simple semigroups. It is shown that if the contracted algebra of a member S of either class is regular then S has to be completely 0-simple (theorems 2.3 and 2.5). In the final section (§3), necessary and sufficient conditions are obtained for the regularity of the contracted algebra of a completely 0-simple semigroup (theorems 3.3 and 3.4).

1. Preliminary results on semigroup algebras

The notation and terminology throughout, with a few minor exceptions, will be that of Clifford and Preston [1961, 1967].

By an *algebra* V over a field F we shall mean a vector space over F endowed with an associative multiplication and satisfying the conditions

$$\alpha(xy) = (\alpha x)y = x(\alpha y)$$

for all $\alpha \in F$ and all $x, y \in V$. We say that V is *regular* if and only if its multiplicative semigroup is regular. Left, right and two-sided ideals of V are defined in the usual way. In the case where V has an identity element, a left ideal of V is termed *finitely generated* if and only if it is of the form

$$Va_1 + Va_2 + \ldots + Va_k$$

for finitely many elements $a_1, a_2, \ldots, a_k$ of V.

The following lemma comprises two results due to von Neumann [1936].

LEMMA 1.1. *Let V be a regular algebra with an identity.*

(i) If M is a finitely generated left ideal of V such that $M \neq V$ then there exists $x \in V\backslash 0$ such that $Mx = 0$.

(ii) If P is an $r \times s$ matrix over V for some $r, s \in \mathbb{N}$ then there exists an $s \times r$ matrix Q over V such that $PQP = P$.

Part (ii) is a minor extension of the corresponding result for square matrices. For a short and simple proof of the latter, see Kaplansky [1967a, theorem 24].

Let S be a semigroup and F a field. By the *semigroup algebra* FS of S over F we mean the set of all mappings from S to F of finite support, with pointwise addition, pointwise scalar multiplication and convolution for multiplication [Clifford and Preston, 1961, §5.2]. As is customary, we consider FS as a vector space over F with S as a basis. The product of two elements of FS is computed using linearity and the multiplication in S in the usual way. For the case in which S is a group, FS is just the *group algebra* (or *group ring*) of S over F.

Now suppose that S has a zero z and that $|S| \geqslant 2$. The subset $Fz = \{\alpha z\colon \alpha \in F\}$ of FS is an ideal of FS and we may consider the factor algebra FS/Fz. This is called the *contracted semigroup algebra* of S over F and it will be denoted here by F_0S. In forming F_0S from FS we have essentially identified the zero of S with that of FS: accordingly in this context we shall henceforth denote the zero of S simply by 0. It is convenient to regard F_0S as a vector space over F with $S\backslash 0$ as a basis; hence a typical nonzero element of F_0S can be written in the form

$$\sum_{i=1}^{m} \alpha_i x_i ,$$

where $x_1, x_2, \ldots, x_m$ are distinct elements of $S\backslash 0$, for some $m \in \mathbb{N}$, and $\alpha_1, \alpha_2, \ldots, \alpha_m$ are nonzero elements of F.

Recall that a group G is said to be *locally finite* if and only if each nonempty finite subset of G generates a finite subgroup of G. Necessary and sufficient conditions for the regularity of a group algebra were provided by Villamayor [1959] and Connell [1963] (see also Passman [1977, p.69]): for ease of reference we state the result as

LEMMA 1.2. *Let F be a field and G a group. Then FG is regular if* *and only if G is locally finite and the characteristic of F is zero or a prime which is not the period of any element of G.*

We also note the following lemma which gives necessary conditions for the regularity of a contracted semigroup algebra [Weissglass 1970c, lemma 2(i) and corollary to lemma 5].

LEMMA 1.3. *Let F be a field and S a semigroup such that $S = S^0$. If $F_0 S$ is regular then*

(i) S is regular;

(ii) every subgroup of S is locally finite and the characteristic of F is zero or a prime which is not the period of any element in a subgroup of S.

2. Regular 0–simple semigroups

In this section we consider two classes of regular 0-simple semigroups. It is shown that if S is a semigroup in either of these classes such that the contracted algebra of S over a field is regular then S must be completely 0-simple (theorems 2.3 and 2.5).

It will be convenient to use the term *monoid* for a semigroup having an identity element. Unless explicit mention is made to the contrary, the identity of a monoid will be denoted by 1.

LEMMA 2.1. *Let S be a regular 0-simple monoid.*

(i) For all $b \in S\backslash 0$ there exists $a \in R_1$ such that $L_a \leq L_b$.

(ii) If $R_1 = H_1$ then S is a group with zero.

Proof. (i) This is a special case of lemma 1.1 of [Munn, 1970a].

(ii) Suppose that $R_1 = H_1$. To show that S is a group with zero it will be enough to prove that 1 is the only idempotent in $S\backslash 0$. Let e be an idempotent in $S\backslash 0$. Then, by (i), there exists $a \in R_1$ such that $L_a \leq L_e$. But $L_a = L_1$, since $R_1 = H_1 \subseteq L_1$. Hence $L_e = L_1$, from which it follows that $e = 1$.

In an arbitrary monoid S the R-class R_1 is a right cancellative subsemigroup, called the *right unit subsemigroup of S*. We begin by considering a regular 0-simple monoid S in which R_1 is left cancellative as well as right cancellative.

LEMMA 2.2. *Let F be a field and S a regular 0-simple monoid whose right unit subsemigroup is (left) cancellative. If $F_{\circ}S$ is regular then S is a group with zero.*

Proof. Let $F_{\circ}S$ be regular. Suppose that $R_1 \neq H_1$. We show that this leads to a contradiction.

Let $p \in R_1 \backslash H_1$. Then there exists $x \in F_{\circ}S$ such that $(1-p)x(1-p) = 1 - p$ and so $[(1-p)x - 1](1-p) = 0$. Hence either (a) $(1-p)x = 1$ or (b) there exists $y \in F_{\circ}S \backslash 0$ such that $y(1-p) = 0$.

Assume that (a) holds. There exist distinct elements $x_1, x_2, \ldots, x_m \in S \backslash 0$ and nonzero scalars $\xi_1, \xi_2, \ldots, \xi_m$ such that $x = \sum_{i=1}^{m} \xi_i x_i$. Now

$$(1 - p^{2m+1})x = 1 + p + p^2 + \ldots + p^{2m}. \tag{1}$$

But the $2m + 1$ elements $1, p, p^2, \ldots, p^{2m}$ of $S \backslash 0$ are distinct (since R_1 is right cancellative and $p \notin H_1$) and are thus linearly independent as elements of $F_{\circ}S$, while the set

$$\{x_1, x_2, \ldots, x_m, p^{2m+1}x_1, p^{2m+1}x_2, \ldots, p^{2m+1}x_m\}$$

contains at most $2m$ distinct elements of $S \backslash 0$. This contradicts equation (1).

Assume that (b) holds. There exist distinct elements $y_1, y_2, \ldots, y_n \in S \backslash 0$ and nonzero scalars $\eta_1, \eta_2, \ldots, \eta_n$ such that $y = \sum_{i=1}^{n} \eta_i y_i$. It follows that

$$\sum_{i=1}^{n} \eta_i y_i = \sum_{i=1}^{n} \eta_i (y_i p).$$

Hence, since the elements of $S \backslash 0$ are linearly independent as elements of $F_{\circ}S$, the mapping

$$y_i \mapsto y_i p \quad (i = 1, 2, \ldots, n)$$

must be a permutation of $\{y_1, y_2, \ldots, y_n\}$. Thus, in particular, there exists a positive integer k such that

$$y_1 = y_1 p^k. \tag{2}$$

But, since S is regular and 0-simple, there exists $a \in R_1$ such that $L_a \leqslant L_{y_1}$, by lemma 2.1(i). Let y_1' denote an inverse of y_1 in S. Then $L_{y_1} = L_{y_1'y_1}$ and so $a = ay_1'y_1$. Hence, from (2), $a = ay_1'y_1p^k$, which shows that $a = ap^k$. Since $a, p^k \in R_1$ and R_1 is left cancellative we deduce that $1 = p^k$. But this implies that $p \in H_1$, contrary to our initial assumption.

Thus $R_1 = H_1$. Consequently, by lemma 2.1(ii), S is a group with zero.

From this result it follows, for example, that if S is the polycyclic monoid $P(I)$ on a nonempty set I, in the sense of Nivat and Perrot [1970b], and F is an arbitrary field then F_0S is not regular: for $P(I)$ is a 0-bisimple inverse monoid whose right unit subsemigroup is the free monoid on I. In particular, by taking $|I| = 1$, we see that the algebra of the bicyclic semigroup B over a field F is not regular, a fact noted by Weissglass [1970c] (for the case $F = \mathbb{Q}$). We now establish this by a direct argument. First, we remark that there exists a homomorphism from B onto the infinite cyclic group C_∞. By linearity, this extends to a homomorphism from FB onto FC_∞. But, by lemma 1.2, FC_∞ is not regular: hence FB is not regular.

A congruence ρ on a semigroup S is termed *idempotent-separating* if and only if no two distinct idempotents of S lie in the same ρ-class. Lallement [1966a] has shown that, on a regular semigroup, a congruence is idempotent-separating if and only if it is contained in $\mathcal{H}$. Consequently a regular semigroup possesses a greatest idempotent-separating congruence μ, namely the greatest congruence contained in $\mathcal{H}$.

We now extend the result of lemma 2.2 to obtain

THEOREM 2.3. *Let F be a field and S a regular 0-simple semigroup. Suppose that, for some nonzero idempotent e in S and some idempotent-separating congruence ρ on eSe, the right unit subsemigroup of eSe/ρ is left cancellative. If F_0S is regular then S is completely 0-simple.*

Proof. We note first that eSe is a regular subsemigroup of S with identity e. Further, since S is 0-simple so also is eSe, as can readily be deduced from lemma 2.28 of [Clifford and Preston 1961].

Hence, since ρ is not the universal congruence, eSe/ρ is a regular 0-simple monoid.

Suppose that F_0S is regular. Then the subalgebra $e(F_0S)e$ of F_0S is also regular. But

$$e(F_0S)e = F_0(eSe)$$

and so $F_0(eSe)$ is regular. Now the canonical homomorphism $\rho^{\natural}: eSe \to eSe/\rho$ defined by $x\rho^{\natural} = x\rho$ $(x \in eSe)$ extends by linearity to a surjective homomorphism from $F_0(eSe)$ to $F_0(eSe/\rho)$. Hence $F_0(eSe/\rho)$ is regular. It follows from lemma 2.2 that eSe/ρ is a group with zero. In particular, eSe/ρ has exactly two idempotents. Thus, since ρ is idempotent-separating, eSe itself has exactly two idempotents, namely e and 0. This shows that e is a primitive idempotent in S and so S is completely 0-simple.

As an illustration of theorem 2.3, consider a 0-simple inverse semigroup S whose semilattice E is an ω-tree with zero (that is, a semilattice each of whose nonzero principal ideals has order type dual to $\omega + 1$). Such semigroups were studied by Maclean [1972]. Let e be a nonzero idempotent of E and let μ denote the greatest idempotent-separating congruence on eSe. Then μ coincides with Green's relation H on eSe, eSe/μ is a combinatorial 0-simple inverse monoid with semilattice of order type dual to $\omega + 1$ and the right unit subsemigroup of eSe/μ is the free monoid on one generator. By theorem 2.3, F_0S is not regular, where F denotes an arbitrary field.

It should be noted that while the right unit subsemigroup of eSe/μ is left cancellative, the right unit subsemigroup of eSe itself need not be left cancellative. This can be seen by taking any group G with a noninjective endomorphism α and considering $S = T^0$, where T is the bisimple ω-semigroup determined by G and α [Reilly, 1966b].

A result similar to lemma 2.2 can be obtained if we replace the hypothesis that the right unit subsemigroup of S is left cancellative by the hypothesis that it is finitely generated. We have

LEMMA 2.4. *Let F be a field and S a regular 0-simple monoid whose right unit subsemigroup is finitely generated. If F_0S is regular then S is a group with zero.*

Proof. Let F_0S be regular and let $p_1, p_2, \ldots, p_m \in R_1$ be such that

$R_1 = \langle p_1, p_2, \ldots, p_m \rangle$. Consider the left ideal M of $F_\circ S$ defined by

$$M = F_\circ S(1 - p_1) + F_\circ S(1 - p_2) + \ldots + F_\circ S(1 - p_m).$$

We show first that $M \neq F_\circ S$. For $i = 1,2,\ldots,m$ there exists $p_i' \in S$ such that $p_i p_i' = 1$: hence, for all $s \in S$,

$$sp_i = 0 \Longleftrightarrow s = 0 \quad (i = 1,2,\ldots,m).$$

It follows that the sum of the coefficients in each element of M must be 0. Hence $M \neq F_\circ S$. But M is finitely generated. Thus, by lemma 1.1(i), there exists $x \in F_\circ S \backslash 0$ such that $Mx = 0$. Since $x \neq 0$ there exist distinct elements $x_1, x_2, \ldots, x_n \in S \backslash 0$ and nonzero scalars $\xi_1, \xi_2, \ldots, \xi_n$ such that $x = \sum_{j=1}^{n} \xi_j x_j$. In particular, $(1 - p_i)x = 0$ $(i = 1,2,\ldots,m)$ and so

$$\sum_{j=1}^{n} \xi_j x_j = \sum_{j=1}^{n} \xi_j (p_i x_j) \quad (i = 1,2,\ldots,m).$$

But the elements of $S \backslash 0$ are linearly independent as elements of $F_\circ S$ and so, for each i, the mapping

$$x_j \mapsto p_i x_j \quad (j = 1,2,\ldots,n) \tag{1}$$

is a permutation of $\{x_1, x_2, \ldots, x_n\}$.

Let x_1' denote an inverse of x_1 in S. Then, by lemma 2.1(i), there exists $a \in R_1$ such that $L_a \leqslant L_{x_1 x_1'}$ and so

$$a = ax_1 x_1'. \tag{2}$$

Now $a = a_1 a_2 \ldots a_r$, where each a_i lies in $\{p_1, p_2, \ldots, p_m\}$. Hence the mapping $x_j \mapsto ax_j$ $(j = 1,2,\ldots,n)$ is a permutation of $\{x_1, x_2, \ldots, x_n\}$. Consequently, there exists a positive integer k such that

$$x_1 = a^k x_1. \tag{3}$$

From (2) and (3) we have that $a^k = a^k x_1 x_1' = x_1 x_1'$. Hence $x_1 \in R_1$, since $a^k \in R_1$. But, from (1), there exist positive integers ℓ_i such that

$$p_i^{\ell_i} x_1 = x_1 \quad (i = 1,2,\ldots,m)$$

and so, since R_1 is right cancellative, $p_i^{\ell_i} = 1$ $(i = 1,2,\ldots,m)$. Thus

$p_i \in H_1$ $(i = 1,2,\ldots,m)$ and so $\langle p_1,p_2,\ldots,p_m\rangle \subseteq H_1$. Since $R_1 = \langle p_1,p_2,\ldots,p_m\rangle$ we have shown that $R_1 = H_1$. It follows from lemma 2.1(ii) that S is a group with zero.

By analogy with theorem 2.3 we now have

THEOREM 2.5. *Let F be a field and S a regular 0-simple semigroup. Suppose that, for some nonzero idempotent e in S and some idempotent-separating congruence ρ on eSe, the right unit subsemigroup of eSe/ρ is finitely generated. If $F_\circ S$ is regular then S is completely 0-simple.*

This follows from lemma 2.4 exactly as theorem 2.3 follows from lemma 2.2. We omit the details.

Finally, we remark that the proof of lemma 2.4 remains valid if we replace the hypothesis that R_1 is finitely generated by the weaker hypothesis that R_1 contains a finitely generated subsemigroup T which meets every left ideal of R_1.

3. Completely 0–simple semigroups

In view of the results in §2 we now focus attention on completely 0-simple semigroups. By the main theorem in [Rees 1940], it suffices to consider the contracted algebra of a Rees matrix semigroup over a group with zero. Algebras of this type have been considered by Weissglass [1969b]: we restate some basic definitions below.

Let F be a field, let G be a group and let I,Λ be arbitrary nonempty sets. By an $I \times \Lambda$ *matrix over* FG we mean a mapping A from $I \times \Lambda$ to FG. The image of (i,λ) under A is denoted by $a_{i\lambda}$ and called the (i,λ)-*entry* of A $(i \in I,\ \lambda \in \Lambda)$. We write $A = (a_{i\lambda})$. The terms 'row', 'column' and 'submatrix' of A will have their obvious meaning. For all nonempty subsets J and M of I and Λ, respectively, the (J,M)-submatrix of A will be denoted by

$$A_{JM}.$$

The $I \times \Lambda$ matrix with all entries 0 will be written simply as 0.

We say that A is *row finite* if and only if, for each $i \in I$, there exists a finite subset Λ_i of Λ such that $a_{i\lambda} = 0$ for all $\lambda \in \Lambda\setminus\Lambda_i$ and we say that A is *row bounded* if and only if there exists a finite subset M of Λ such that $a_{i\lambda} = 0$ for all $i \in I$ and all $\lambda \in \Lambda \backslash M$. The *column support*, col supp A, of A is defined by

$$\text{col supp } A = \{\lambda \in \Lambda \colon a_{i\lambda} \neq 0 \text{ for some } i \in I\}.$$

Clearly A is row bounded if and only if its column support is finite. Column finiteness, column boundedness and the row support, row supp A, of A are defined in a dual fashion. We say that A is *bounded* if and only if it is both row bounded and column bounded. The set of all bounded $I \times \Lambda$ matrices over FG forms a vector space with respect to pointwise addition and multiplication by scalars.

Let $A = (a_{i\lambda})$ be an $I \times \Lambda$ matrix over FG and $B = (b_{\lambda j})$ a $\Lambda \times J$ matrix over FG, where J is another nonempty set. For all $(i,j) \in I \times J$ let

$$S_{ij} = \{\lambda \in \Lambda \colon a_{i\lambda} b_{\lambda j} \neq 0\}.$$

Then the product AB is defined if and only if each S_{ij} is finite: if each S_{ij} is finite we take AB to be the $I \times J$ matrix over FG whose (i,j)-entry is

$$\sum_{\lambda \in S_{ij}} a_{i\lambda} b_{\lambda j}$$

if $S_{ij} \neq \emptyset$ and 0 if $S_{ij} = \emptyset$. Now suppose that C is a $J \times M$ matrix over FG, for some nonempty set M, and that the products AB, BC, $(AB)C$ and $A(BC)$ are defined. Then it can be verified that $(AB)C = A(BC)$. Accordingly we may omit brackets and write either of the last two products simply as ABC. Note further that if A and C are bounded then so also is ABC.

For any nonempty set I the $I \times I$ unity matrix over FG will be denoted by Δ_I.

An $I \times \Lambda$ matrix A over FG is *left* [*right*] *invertible* (*over FG*) if and only if there exists a $\Lambda \times I$ matrix B over FG such that $BA = \Delta_\Lambda$ [$AB = \Delta_I$]. We say that A is *invertible* (*over FG*) if and only if it is both left and right invertible.

It is easy to verify that an $I \times \Lambda$ matrix A over FG is invertible if and only if there exists a unique $\Lambda \times I$ matrix B over FG such that

$$AB = \Delta_I, \quad BA = \Delta_\Lambda.$$

If A is invertible we denote B above by A^{-1}.

Note that we have not assumed here that $|I| = |\Lambda|$.

The first lemma of this section concerns matrices of finite type.

LEMMA 3.1. *Let F be a field, let G be a locally finite group, let M and J be nonempty finite sets and let Q be a left invertible $M \times J$ matrix over FG. Then*

(i) $|M| \geqslant |J|$,

(ii) $|M| = |J| \Rightarrow Q$ *is invertible.*

Proof. By hypothesis, there exists a $J \times M$ matrix R over FG such that $RQ = \Delta_J$. Let H denote the subgroup of G generated by those elements of G which appear with nonzero coefficients in the entries of Q or R. Since M and J are finite and G is locally finite, H is a finite subgroup of G. We may therefore regard Q and R as matrices over the finite-dimensional group algebra FH. Let ρ denote the regular representation of FH by matrices over F and $Q\rho, R\rho$ denote the matrices over F obtained from Q, R respectively by replacing each entry by its image under ρ. Then $(R\rho)(Q\rho) = \Delta_J\rho$ and so, by considering the rank of $Q\rho$, we see that $|M||H| \geqslant |J||H|$. Hence $|M| \geqslant |J|$.

Next, suppose that $|M| = |J|$. Without loss of generality we assume that $M = J$. With R and ρ as before we have that $(Q\rho)(R\rho) = \Delta_J\rho$, since $\Delta_J\rho$ is the $|J||H| \times |J||H|$ unity matrix over F; that is, $(QR)\rho = \Delta_J\rho$. Hence, since ρ is a faithful representation of FH, $QR = \Delta_J$. Thus Q is invertible.

Now consider the Rees matrix semigroup $S = M^0(G;I,\Lambda;P)$, where G is a group, I and Λ are nonempty sets and P is a $\Lambda \times I$ matrix over G^0. By lemma 3.1 of [Clifford and Preston 1961], S is regular if and only if each row and each column of P has a nonzero entry. If S is regular then it is completely 0-simple; conversely, every completely 0-simple semigroup is isomorphic to a regular semigroup S of this form [Rees 1940; see also Clifford and Preston 1961, theorem 3.5]. It is readily verified that, in the completely 0-simple case, S has finitely many right ideals if and only if $|I| < \infty$ and finitely many left ideals if and only if $|\Lambda| < \infty$.

We can regard P as a $\Lambda \times I$ matrix over FG. As already observed, the set V of all bounded $I \times \Lambda$ matrices over FG forms a vector space under pointwise addition and multiplication by scalars. Now define a multiplication ($\circ$) on V by the rule that

$$A \circ B = APB \quad (A, B \in V).$$

With respect to this multiplication V becomes an algebra over F: we denote it by $M(FG; I, \Lambda; P)$. As remarked by Weissglass [1969b, lemma 3.1],

$$F_0 S \cong M(FG; I, \Lambda; P).$$

For simplicity we shall henceforth identify $F_0 S$ with $M(FG; I, \Lambda; P)$.

LEMMA 3.2. *Let F be a field, let $S = M^0(G; I, \Lambda; P)$ be a regular Rees matrix semigroup over a group with zero G^0 and let $F_0 S$ be regular. Let J be a nonempty finite subset of I such that $|J| \leqslant |\Lambda|$. Then there exists a nonempty finite subset M of Λ such that P_{MJ} is left invertible over FG.*

Proof. Let $\theta\colon J \to \Lambda$ be an injective mapping, let N denote $J\theta$ and let $U = (u_{j\lambda})$ be the $J \times N$ matrix over FG defined by

$$u_{j\lambda} = \begin{cases} \text{the identity of } G \text{ if } \lambda = j\theta, \\ 0 \text{ otherwise.} \end{cases} \qquad (\lambda \in \Lambda, j \in J)$$

Then U is invertible and $UU^{-1} = \Delta_J$. Now define $A \in F_0 S = M(FG; I, \Lambda; P)$ by

$$\text{row supp } A = J, \text{ col supp } A = N, \; A_{JN} = U.$$

Since $F_0 S$ is regular, there exists $X \in F_0 S \backslash 0$ such that $APXPA = A$. Write

$$K = \text{row supp } X, \; M = \text{col supp } X.$$

Then

$$A_{JN} P_{NK} X_{KM} P_{MJ} A_{JN} = A_{JN}.$$

Hence, since $A_{JN} = U$,

$$(UP_{NK} X_{KM}) P_{MJ} = UU^{-1} = \Delta_J.$$

Thus P_{MJ} is left invertible.

We now establish the first main result of this section.

THEOREM 3.3. *Let F be a field and let $S = M^0(G;I,\Lambda;P)$ be a regular Rees matrix semigroup over a group with zero G^0, with at least one of the sets I and Λ finite. Then F_0S is regular if and only if the following conditions hold:*

(i) G is locally finite,

(ii) the characteristic of F is zero or a prime which is not the period of any element of G,

(iii) $|I| = |\Lambda|$ and P is invertible over FG.

Proof. Assume first that F_0S is regular. Since S has a subgroup isomorphic to G it follows from lemma 1.3(ii) that (i) and (ii) hold.

Without loss of generality we suppose that Λ is finite and that $|\Lambda| \leqslant |I|$. Let J be a finite subset of I such that $|J| = |\Lambda|$. Then, by lemma 3.2, there exists a nonempty subset M of Λ such that P_{MJ} is left invertible over FG. But $|J| \geqslant |M|$ and, by lemma 3.1(i), $|M| \geqslant |J|$. Hence $|M| = |J|$ and so $M = \Lambda$. Thus, by lemma 3.1(ii), $P_{\Lambda J}$ is invertible. Now suppose that $|\Lambda| < |I|$: we shall show that this leads to a contradiction. Define an $I \times I$ matrix Q over FG as follows:

$$\begin{cases} Q_{JJ} = \Delta_J, & Q_{J,I\setminus J} = P_{\Lambda J}^{-1}P_{\Lambda,I\setminus J}, \\ Q_{I\setminus J,J} = 0, & Q_{I\setminus J,I\setminus J} = \Delta_{I\setminus J}. \end{cases}$$

Thus, in the notation of partitioned matrices,

$$Q = \begin{bmatrix} \Delta_J & P_{\Lambda J}^{-1}P_{\Lambda,I\setminus J} \\ 0 & \Delta_{I\setminus J} \end{bmatrix} .$$

Now Q is column finite and invertible: indeed, with the same notation,

$$Q^{-1} = \begin{bmatrix} \Delta_J & -P_{\Lambda J}^{-1}P_{\Lambda,I\setminus J} \\ 0 & \Delta_{I\setminus J} \end{bmatrix} .$$

Define a $\Lambda \times I$ matrix $\hat{P}$ over G^0 by

$$\hat{P}_{\Lambda J} = P_{\Lambda J}, \quad \hat{P}_{\Lambda,I\setminus J} = 0.$$

Then

$$\hat{P}Q = P. \tag{1}$$

Write $\hat{S} = M^0(G;I,\Lambda;\hat{P})$ and define $\theta\colon F_\circ S \to F_\circ \hat{S}$ by

$$X\theta = QX \quad (X \in F_\circ S).$$

(Note that if X is a bounded $I \times \Lambda$ matrix over FG then so also is QX for Q as above.) Then θ is linear and, since Q is invertible, θ is a bijection. Also, by (1),

$$Q(XPY) = Q(X(\hat{P}Q)Y) = (QX)\hat{P}(QY)$$

for all $X,Y \in F_\circ S$. Hence θ is an algebra isomorphism. But $\hat{S}$ is not a regular semigroup, since $\hat{P}$ has at least one zero column, and so, by lemma 1.3(i), $F_\circ \hat{S}$ cannot be regular. Hence $F_\circ S$ is not regular and this contradicts our hypothesis. Consequently $|\Lambda| = |I|$. It follows that $J = I$ and hence that $P(=P_{\Lambda J})$ is invertible. Thus (iii) holds.

Now assume, conversely, that (i), (ii) and (iii) hold. Write $T = M^0(G;I,I;\Delta_I)$ and define $\phi\colon F_\circ S \to F_\circ T$ by

$$X\phi = PX \quad (X \in F_\circ S).$$

Since (iii) holds, ϕ is an algebra isomorphism. Also, since (i) and (ii) hold, FG is regular, by lemma 1.2. Hence, by lemma 1.1(ii), $F_\circ T$ is regular. Thus $F_\circ S$ is regular.

Theorem 3.3 generalises an earlier result (Munn [1955], Ponizovskiǐ [1956]) giving necessary and sufficient conditions for the semisimplicity of the contracted algebra over F of a *finite* 0-simple semigroup.

We turn now to the case in which the sets I and Λ are both infinite. Here we require a further definition.

Let P be a $\Lambda \times I$ matrix over FG. Then P is *locally left invertible* (*over FG*) if and only if to each nonempty finite subset J of I there corresponds a nonempty finite subset M of Λ such that P_{MJ} is left invertible. Similarly, P is *locally right invertible* (*over FG*) if and only if to each nonempty finite subset M of Λ there corresponds a nonempty finite subset J of I such that P_{MJ} is right invertible.

The second main result follows.

THEOREM 3.4. *Let F be a field and let $S = M^0(G;I,\Lambda;P)$ be a regular Rees matrix semigroup over a group with zero, with I and Λ infinite. Then F_0S is regular if and only if the following conditions hold:*

(i) G is locally finite,

(ii) the characteristic of F is zero or a prime which is not the period of any element of G,

(iii) P is locally left and right invertible over FG.

Proof. Assume first that F_0S is regular. Since S has a subgroup isomorphic to G it follows from lemma 1.3(ii) that (i) and (ii) hold.

Let J be a nonempty finite subset of I. Then $|J| < |\Lambda|$ (since Λ is infinite) and so, by lemma 3.2, there exists a nonempty finite subset M of Λ such that P_{MJ} is left invertible over FG. Thus P is locally left invertible over FG. A similar argument shows that P is locally right invertible. Hence (iii) holds.

Assume, conversely, that (i), (ii) and (iii) hold. Let $A \in F_0S\backslash 0$. Write

$$J = \text{row supp } A, \qquad M = \text{col supp } A.$$

Since (iii) holds, there exist nonempty finite subsets K of I and N of Λ such that P_{MK} is right invertible and P_{NJ} left invertible over FG. Thus there exists a $K \times M$ matrix Q over FG and a $J \times N$ matrix R over FG such that

$$P_{MK}Q = \Delta_M, \qquad RP_{NJ} = \Delta_J. \tag{1}$$

Since (i) and (ii) hold, FG is regular, by lemma 1.2. Hence, by lemma 1.1(ii), there exists an $M \times J$ matrix Y over FG such that $A_{JM}YA_{JM} = A_{JM}$. Thus, from (1),

$$A_{JM} = A_{JM}P_{MK}(QYR)P_{NJ}A_{JM}.$$

Now define X to be the $I \times \Lambda$ matrix over FG such that

$$\text{row supp } X \subseteq K, \quad \text{col supp } X \subseteq N, \quad X_{KN} = QYR.$$

Then

$$(APXPA)_{JM} = A_{JM}P_{MK}X_{KN}P_{NJ}A_{JM} = A_{JM}$$

and so $APXPA = A$. This shows that F_0S is regular.

We conclude with some examples to illustrate theorems 3.3 and 3.4. As before, F denotes a field.

Let G be a group, let I be a nonempty set and let S denote the Brandt semigroup $M^0(G;I,I;\Delta_I)$. Then, by theorems 3.3 and 3.4, F_0S is regular if and only if G is locally finite and the characteristic of F is zero or a prime which is not the period of any element of G. This can also be deduced from theorems 1 and 2 of [Weissglass 1970c]. We can generalise the result as follows. Suppose that I is totally ordered and let $P = (p_{ij})$ be any $I \times I$ matrix over G^0 such that

$$\begin{cases} p_{ii} \neq 0 & \text{for all } i \in I, \\ p_{ij} = 0 & \text{for all } (i,j) \in I \times I \text{ such that } i > j. \end{cases}$$

Take $S = M^0(G;I,I;P)$. Then, for each nonempty finite subset J of I, P_{JJ} is a triangular matrix over FG with nonzero entries on the main diagonal and is therefore invertible. Thus, by theorems 3.3 and 3.4, F_0S is regular if and only if G is locally finite and the characteristic of F is zero or a prime which is not the period of any element of G. For suitably chosen I and G, one can construct matrices P of the above type which are not themselves invertible.

Another example shows that the choice of F may be important even when the structure group of the completely 0-simple semigroup is trivial. Take I to be a nonempty set, define an $I \times I$ matrix $P = (p_{ij})$ over F (with unity 1) by

$$\begin{cases} p_{ii} = 0 \text{ for all } i \in I, \\ p_{ij} = 1 \text{ for all } (i,j) \in I \times I \text{ with } i \neq j, \end{cases}$$

and let $S = M^0(1;I,I;P)$. We separate two cases.

(a) Suppose that I is finite. It is easily verified that P is invertible over $F(=F1)$ if and only if F has characteristic zero or a prime not dividing $|I| - 1$. Thus, by theorem 3.3, F_0S is regular if and only if F has characteristic zero or a prime not dividing $|I| - 1$.

(b) Suppose that I is infinite. For each $n \in \mathbb{N}$ let A_n denote the $n \times n$ matrix over F whose (i,j)-entry is

$$\begin{cases} 0 \text{ if } j = i - 1 \ (i = 2,3,\ldots,n), \\ 1 \text{ otherwise.} \end{cases}$$

Then A_n is invertible over F (whatever the characteristic of F). It follows readily that P is locally left and right invertible. Thus, by theorem 3.4, F_0S is regular.

Finally, we remark that if $I = \mathbb{N}$ in this example then P itself is not invertible over F. To see this, suppose that $I = \mathbb{N}$ and that $Q = (q_{ij})$ is an $\mathbb{N} \times \mathbb{N}$ matrix over F such that $PQ = \Delta_{\mathbb{N}}$. Evidently Q must be column finite. Let $m \in \mathbb{N}$ be such that $q_{m1} \neq 0$ and $q_{i1} = 0$ for all $i > m$. Then $m \geqslant 2$ and so, by considering the $(m,1)$- and $(m+1,1)$-entries of PQ, we find that

$$q_{11} + q_{21} + \ldots + q_{m-1,1} = 0 = q_{11} + q_{21} + \ldots + q_{m1}.$$

Thus $q_{m1} = 0$, which contradicts the definition of m. Consequently no such matrix Q exists.

References

1936 J. von Neumann, On regular rings. *Proc. Nat. Acad. Sci. U.S.A.*, 22 (1936), 707-13.

1940 D. Rees, On semi-groups. *Proc. Cambridge Philos. Soc.*, 36 (1940), 387-400.

1955 W.D. Munn, On semigroup algebras. *Proc. Cambridge Philos. Soc.*, 51 (1955), 1-15.

1956 I.S. Ponizovskiĭ, On matrix representations of associative systems. *Mat. Sbornik*, 38 (1956), 241-60. [In Russian.]

1959 O.E. Villamayor, On weak dimension of algebras. *Pacific J. Math.*, 9 (1959), 941-51.

1961 A.H. Clifford and G.B. Preston, *The algebraic theory of semi-groups, Vol.1.* Math. Surveys of the Amer. Math. Soc. 7, 1961.

1963 I.G. Connell, On the group ring. *Canad. J. Math.*, 15 (1963), 650-85.

1966a G. Lallement, Congruences et équivalences de Green sur un demi-groupe régulier. *C.R. Acad. Sci. Paris (Sér. A)*, 262 (1966), 613-6.

1966b N.R. Reilly, Bisimple ω-semigroups. *Proc. Glasgow Math. Assoc.*, 7 (1966), 160-7.

1967 A.H. Clifford and G.B. Preston, *The algebraic theory of semigroups, Vol. 2.* Math. Surveys of the Amer. Math. Soc. 7, 1967.

1969a I. Kaplansky, *Fields and rings.* The University of Chicago Press, 1969.

1969b J. Weissglass, Radicals of semigroup rings. *Glasgow Math. J.*, 10 (1969), 85-93.

1970a W.D. Munn, Fundamental inverse semigroups. *Quarterly J. Math. Oxford (2)*, 21 (1970), 157-70.

1970b M. Nivat and J.-F. Perrot, Une généralisation du monoïde bicyclique. *C.R. Acad. Sci. Paris (Sér. A)*, 271 (1970), 824-7.

1970c J. Weissglass, Regularity of semigroup rings. *Proc. Amer. Math. Soc.*, 25 (1970), 499-503.

1972 P.M. Maclean, *Contributions to the theory of 0-simple inverse semigroups.* Doctoral thesis. The University of Stirling, 1972.

1977 D.S. Passman, *The algebraic structure of group rings.* Wiley-Interscience, 1977.

SEMIGROUPS AND GRAPHS

G. B. Preston

Department of Mathematics, Monash University, Clayton,
Victoria, Australia, 3168.

Abstract

Various semigroups are associated with a graph. We start with the semigroup of paths, then from this form its maximal inverse semigroup morphic image, the inverse semigroup of the graph. We also consider the free inverse semigroup generated by the arrows of the graph and show that the inverse semigroup of the graph is a Rees quotient of this free inverse semigroup. What is sometimes called the fundamental groupoid of the graph is identified with the maximal primitive morphic image of the inverse semigroup of the graph.

1. Introduction

The possibility of using graphs effectively in semigroup theory, repeating the successes of group theory, opens up tantalising vistas. The work of W.D. Munn [1973] on free inverse semigroups in which he used graphs effectively to determine canonical forms for the elements of a free inverse semigroup shows how powerful and illuminating graphs can be. Monash University recently had Professor Teh H.H., of the Nanyang University, as a visitor, and during his stay he gave a series of lectures on semigroups and graphs. In particular he listed a number of inverse semigroups related to graphs which he exhibited for exploration. I mention now some of these. For an expanded account see Teh and Shee [1979].

Teh took a graph to be an ordered pair $\langle G,A \rangle$ where $A \subseteq G \times G$. G is the set of *vertices* of $\langle G,A \rangle$ and A is its set of *edges* or *arrows*. Thus the arrows are directed and there is at most one arrow from any vertex to any other vertex: we say that the arrow (a,b) is *from* the vertex a *to* the vertex b. The vertex a is called the *origin* of (a,b)

ISBN 0-12-319450-4

and the vertex b is called its *terminus*.

Let $\langle G,A\rangle$ and $\langle H,B\rangle$ be graphs. A *partial isomorphism* of $\langle G,A\rangle$ into $\langle H,B\rangle$ is a mapping $\alpha: N \to H$, of a subset N of G into H, which is one-to-one and which preserves origins and terminals of arrows, i.e. is such that for all $(a,b) \in A \cap (N \times N)$ we have $(a\alpha, b\alpha) \in B$.

Let I_X denote the symmetric inverse semigroup on a set X. If S and T are inverse semigroups then by $S \leqslant T$ we mean that S is an inverse subsemigroup of T.

Among the examples of inverse semigroups associated with graphs considered by Teh were the following (A), (B), and (C).

(A) *The semigroup* $Q_{\langle G,A\rangle}$ *of all partial automorphisms of* $\langle G,A\rangle$, *i.e. of all partial isomorphisms of* $\langle G,A\rangle$ *into* $\langle G,A\rangle$.

The semigroup $Q_{\langle G,A\rangle} \leqslant I_G$ and, as Teh showed, any inverse semigroup can be faithfully represented as an inverse subsemigroup of some $Q_{\langle G,A\rangle}$.

(B) *The semigroup* $S_{\langle G,A\rangle} \leqslant I_G$.

$S_{\langle G,A\rangle}$ consists of all α such that

$$\alpha = \rho_1 \circ \rho_2 \circ \dots \circ \rho_k,$$

for some $k \geqslant 1$, $\rho_i \in I_G$ and such that $\rho_i \subseteq A \cup A^{-1}$, for $i = 1,2,\dots,k$.

We can provide a characterization of $S_{\langle G,A\rangle}$. First we need some definitions.

An (undirected) *path* in the graph $\langle G,A\rangle$ is a sequence

$$a_0, a_1, \dots, a_n$$

such that $(a_{i-1}, a_i) \in A \cup A^{-1}$, for $i = 1,2,\dots,n$. Such a path is said to be of *length* n and to be from a_0 to a_n. The vertex a_i is said to be the *i-th step* in this path.

Two paths, one from a to b and the other from c to d are said to be *parallel* if either they are identical or, alternatively, they are of the same length and, for each i, the i-th step from a to b is different from the i-th step from c to d.

THEOREM. $S_{\langle G,A\rangle}$ *consists of all* $\alpha \subseteq G \times G$ *such that, for some* k, *any two elements* (a,b), (c,d) *of* α *are such that there are parallel paths of length* k *from* a *to* b *and from* c *to* d.

Proof. Let $\alpha \in S_{\langle G,A\rangle}$. Then $\alpha = \rho_1 \circ \ldots \circ \rho_k$, say, where each $\rho_i \in I_G$ and $\rho_i \subseteq A \cup A^{-1}$. Thus, if (a,b), (c,d) both belong to α, there are paths $a = a_0, a_1, \ldots, a_k = b$ and $c = c_0, c_1, \ldots, c_k = d$ such that $(a_{i-1}, a_i) \in \rho_i$ and $(c_{i-1}, c_i) \in \rho_i$, for $i = 1,2,\ldots,k$. Because each $\rho_i \in I_G$, if for any i, $a_i = c_i$, then $a_j = c_j$, for $j = 0,1,\ldots,k$. Thus the above paths from a to b and from c to d are parallel.

Conversely, let α satisfy, for the number k, the condition stated in the theorem. Let

$$\alpha = \{(a^{(j)}, b^{(j)}) \mid j \in J\}$$

and let

$$a^{(j)} = a_0^{(j)}, a_1^{(j)}, \ldots, a_k^{(j)} = b^{(j)}, \quad j \in J$$

be the parallel paths, of length k, that exist by assumption. Set

$$\rho_i = \{(a_{i-1}^{(j)}, a_i^{(j)}) \mid j \in J\},$$

for $i = 1,2,\ldots,k$. Then, by assumption, $a_i^{(j)} = a_i^{(\ell)}$ if and only if $a_{i-1}^{(j)} = a_{i-1}^{(\ell)}$. Hence each $\rho_i \in I_G$. Moreover, by the definition of a path, each $(a_{i-1}^{(j)}, a_i^{(j)}) \in A \cup A^{-1}$. Hence $\rho_i \subseteq A \cup A^{-1}$.

Thus $\rho_1 \circ \rho_2 \circ \ldots \circ \rho_k \in S_{\langle G,A\rangle}$. By construction $\alpha \subseteq \rho_1 \circ \rho_2 \circ \ldots \circ \rho_k$.

Conversely, let $(x,y) \in \rho_1 \circ \rho_2 \circ \ldots \circ \rho_k$. Thus there exist $x = x_0, x_1, \ldots, x_k = y$, say, such that $(x_{i-1}, x_i) \in \rho_i$, for $i = 1,2,\ldots,k$. Hence

$$(x_0, x_1) = (a_0^{(j)}, a_1^{(j)}), \text{ for some } j,$$

and

$$(x_1, x_2) = (a_1^{(\ell)}, a_2^{(\ell)}), \text{ for some } \ell.$$

But $x_1 = a_1^{(j)} = a_1^{(\ell)}$, and since paths j and ℓ are parallel, since they co-incide at the first step they must be identical. Thus $j = \ell$. Similarly, $x_3, x_4, \ldots, x_k$ are vertices on the j-th path. Thus $x = a^{(j)}$,

$y = b^{(j)}$ and $(x,y) \in \alpha$, showing that $\rho_1 \circ \rho_2 \circ \dots \circ \rho_k \subseteq \alpha$.

Hence $\alpha = \rho_1 \circ \rho_2 \circ \dots \circ \rho_k$; which completes the proof of the theorem.

(C) *Inverse semigroup graphs.*

Teh defined an *inverse semigroup graph* $\langle G,A,\cdot\rangle$ to be a graph $\langle G,A\rangle$ such that $\langle G,\cdot\rangle$ is an inverse semigroup and is such that

$$\forall\ a,b,c \in G,\ (a,b) \in A \Rightarrow (ca,cb) \in A.$$

CONSTRUCTION. *Let* $\langle G,\cdot\rangle$ *be an inverse semigroup. Let* $A \subseteq G \times G$. *Define*

$$A^* = \{(ca,cb)\,|\,c \in G,\ (a,b) \in A\}.$$

Then $\langle G,A^*,\cdot\rangle$ *is an inverse semigroup graph.*

All inverse semigroup graphs are clearly obtained in this way; for if $\langle G,A,\cdot\rangle$ is an inverse semigroup graph, then $A^* = A$.

As a further example Teh gave: Let $\langle G,\cdot\rangle$ be an inverse semigroup and let $X \subseteq G$. Define

$$A_X = \{(g,gx)\,|\,g \in G,\ x \in X\}.$$

Then (G,A_X) is an inverse semigroup graph.

When the inverse semigroups concerned are groups then the above graphs (G,A_X) become the well-known graphs of a group. An important feature of the group graph is that an arrow (g,gx) is uniquely determined by g and x: if (g,h) is an arrow then $(g,h) = (g,gx)$ for a unique $x = g^{-1}h$. This feature disappears if G is an inverse semigroup or, in general, any semigroup.

This suggests that we should also investigate, in connexion with inverse semigroups, graphs with multiple arrows connecting a pair of vertices. One way of doing this is the concern of the rest of this paper.

2. Preliminary definitions

A graph (see S. Mac Lane [1971]) $\Gamma = (V(\Gamma), A(\Gamma)) = (V, A)$, is an ordered pair of sets, V being the set of *vertices* of Γ and A its set of *arrows*, together with a pair of mappings.

$$A \underset{t}{\overset{o}{\rightrightarrows}} V \; .$$

For $a \in A$, $o(a)$ is called the *origin* and $t(a)$ is called the *terminus* of the arrow a. Together $o(a)$ and $t(a)$ are called the *end points of a*. We shall frequently denote a graph simply by its set of arrows.

We introduce a set A^{-1}, disjoint from A, such that $a \mapsto a^{-1}$, $a \in A$, is a bijection of A upon A^{-1}. Set $(a^{-1})^{-1} = a$, so that $a \mapsto a^{-1}$, $a \in A^{-1}$, is the reverse bijection. Define $o(a^{-1}) = t(a)$ and $t(a^{-1}) = o(a)$, for a in A; whence the same equations hold for $a \in A^{-1}$. This extension of the mappings o and t makes $(V, A \cup A^{-1})$ a graph. We denote this graph by Γ^* and call it the *inverse closure* of Γ.

A *path* p, of *length* k, in Γ^*, from α to β, is a sequence

$$p = a_1 a_2 \ldots a_k,$$

for which $a_i \in A \cup A^{-1}$ and $t(a_i) = o(a_{i+1})$, $i = 1,2,\ldots,k-1$, and where $\alpha = o(a_1)$, $\beta = t(a_k)$. Define $o(p) = \alpha$, the *origin* of the path p, and $t(p) = \beta$, the *terminus* of p. The path p from α to β is said to *connect* α and β.

Denote by $P(\Gamma^*)$ the set of all paths in Γ^*. Then the pair $(V(\Gamma), P(\Gamma^*))$, together with the extensions of o and t to $P(\Gamma^*)$ just defined, is a graph. We shall usually denote this graph simply by $P(\Gamma^*)$.

The relation on vertices of Γ (or Γ^*) such that α and β are in that relation if and only if $\alpha = \beta$ or α and β are connected by a path in Γ^* is an equivalence relation on $V(\Gamma)$. The equivalence classes of this relation form the sets of vertices of the *connected components* of Γ (and of Γ^*). Specifically, if V_i, $i \in I$, are the equivalence classes, set

$$A_i = \{a \in A \,|\, o(a) \in V_i,\ t(a) \in V_i\},$$

then $\Gamma_i = (V_i, A_i)$, $i \in I$, are the connected components of Γ and

$\Gamma_i^* = (V_i, A_i \cup A_i^{-1})$ are the connected components of Γ^*. In each component Γ_i (or Γ_i^*) any two vertices are connected by a path in Γ_i^*.

If $p = a_1 a_2 \ldots a_k$ is a path in Γ^* then we define p^{-1} to be the path $a_k^{-1} \ldots a_2^{-1} a_1^{-1}$. Thus $(p^{-1})^{-1} = p$ and $o(p) = t(p^{-1})$, $t(p) = o(p^{-1})$.

3. The inverse semigroup of a graph

Let $P = P(\Gamma^*)$ be the set of paths in Γ^* and let $P^0(\Gamma^*) = P^0 = P \cup \{0\}$, where $0 \notin P$. Define a product on P^0 by:

$$uv = \begin{cases} \text{the path } uv, \text{ if } u,v \in P, \text{ and } t(u) = o(v), \\ 0, \text{ otherwise.} \end{cases}$$

Then P^0 is a semigroup with zero called the *involutory semigroup* of the graph Γ: if we set $0^{-1} = 0$, then it may be checked that the mapping $p \mapsto p^{-1}$, $p \in P^0$, is an involution, i.e. $(p^{-1})^{-1} = p$ and $(pq)^{-1} = q^{-1}p^{-1}$, for all p,q in P^0.

If we also set $o(0) = t(0) = \theta$, adding an extra vertex θ to $V(\Gamma)$, to give $V^0 = V^0(\Gamma) = V \cup \{\theta\}$, then $P^0(\Gamma^*)$ becomes a graph in which the graph $P(\Gamma^*)$ is embedded and with vertices $V(P^0) = V^0$.

Define the relation $\sim$ on P^0 by

$$\sim \; = \{(yy^{-1}y, y) \mid y \in P^0\} \cup \{(yy^{-1}zz^{-1}, zz^{-1}yy^{-1}) \mid y,z \in P^0\}.$$

Let $\sim^*$ be the congruence on P^0 generated by $\sim$.

LEMMA 3.1. *$P^0/{\sim^*}$ is an inverse semigroup; indeed it is the maximal inverse semigroup morphic image of P^0.*

Proof. The proof is similar to the well-known proof that (see later, §4) $F_{A \cup A^{-1}}/{\sim^*}$ is the free inverse semigroup on A.

DEFINITION. $P^0(\Gamma)/{\sim^*}$ will be denoted by $\kappa(\Gamma)$ and called the *inverse semigroup of the graph* Γ.

LEMMA 3.2. *If $x \in \kappa(\Gamma)$ and $u,v \in x$, then $o(u) = o(v)$ and $t(u) = t(v)$.*

Proof. If $(p,q) \in \sim$ then it is easily checked that $o(p) = o(q)$

and $t(p) = t(q)$. Hence each $\sim$-transition of u preserves the origin and terminus of u.

DEFINITION. If $x \in \kappa(\Gamma)$ then we define $o(x)$ and $t(x)$ by $o(x) = o(u)$ and $t(x) = t(u)$, where $u \in x$. This makes $\kappa(\Gamma)$ into a graph with vertices $V(\kappa(\Gamma)) = V(P^0(\Gamma))$.

DEFINITION. Each vertex v of $V(\kappa(\Gamma))$ determines an inverse subsemigroup, possibly empty, $\kappa(\Gamma,v)$ of $\kappa(\Gamma)$ whose elements are $x \in \kappa(\Gamma)$ such that $o(x) = t(x) = v$.

Define a relation σ on $\kappa(\Gamma)$ thus:

$$x \,\sigma\, y \text{ if and only if } \quad x = y = 0, \text{ the zero of } \kappa(\Gamma), \\ \text{or, } ex = ey \neq 0, \text{ for some } \\ e = e^2 \text{ in } \kappa(\Gamma).$$

The following lemma is straightforward to prove.

LEMMA 3.3. *The relation σ is a congruence on $\kappa(\Gamma)$ and $\kappa(\Gamma)/\sigma$ is an inverse semigroup such that, writing $(\sigma,v) = \sigma|\kappa(\Gamma,v)$ then $\kappa(\Gamma,v)/(\sigma,v)$ is the maximal group morphic image of $\kappa(\Gamma,v)$.*

DEFINITION. We denote $\kappa(\Gamma)/\sigma$ by $\pi(\Gamma)$ and call it the primitive semigroup of Γ. We denote $\kappa(\Gamma,v)/(\sigma,v)$ by $\pi(\Gamma,v)$.

LEMMA 3.4. *If $u \in \pi(\Gamma)$ and $x,y \in u$ then $t(x) = t(y)$ and $o(x) = o(y)$.*

Proof. By definition of σ, $x,y \in u$ if and only if $x = y = 0$, when $t(x) = t(y) = o(x) = o(y) = \theta$, or there exists $e = e^2$ such that $ex = ey \neq 0$. In the latter event, $o(x) = o(y) = o(e)$ and $t(x) = t(y) = t(ex)$.

This lemma permits the following

DEFINITION. Let $u \in \pi(\Gamma)$. Define $o(u) = o(x)$ and $t(u) = t(x)$, where $x \in u$. $\pi(\Gamma)$ thereby becomes a graph and $V(\pi(\Gamma)) = V^0(\Gamma)$.

LEMMA 3.5. *The natural mappings*

$$\Gamma \to \Gamma^* \to P^0(\Gamma^*) \to \kappa(\Gamma) \to \pi(\Gamma)$$

are graph morphisms and Γ is successively embedded in each of these graphs by these mappings.

Proof. Arrows in Γ are embedded as paths in $P^0(\Gamma)$ and preserve their origins and termini. Hence the first two mappings are graph embeddings. Similarly, if $p \in A(P^0(\Gamma^*))$, then $o(p) = o(p\sim^*) = o(p\sim^*\sigma)$ and $t(p) = t(p\sim^*) = t(p\sim^*\sigma)$. Hence each of the mappings exhibited is a graph morphism.

It remains to show that Γ is embedded, successively, in $\kappa(\Gamma)$ and $\pi(\Gamma)$.

Let $y,z \in A(\Gamma)$. Then, regarding y and z as elements of $P(\Gamma^*)$, $y\sim^*z$ only if $y = z$, by the definition of $\sim$, and $y\sim^*\sigma = z\sim^*\sigma$ means there exists $e = e^2 \in \kappa(\Gamma)$, since $y\sim^*$ and $z\sim^*$ are both $\neq 0$, such that $e(y\sim^*) = e(z\sim^*) \neq 0$.

Since $e = ee^{-1}$, there exists $p \in P(\Gamma^*)$ such that $pp^{-1} \in e$ and $pp^{-1}y\sim^*pp^{-1}z$.

For each arrow $a \in A \cup A^{-1}$ that occurs in p its inverse a^{-1} occurs in p^{-1}. Hence the sum of the number of occurrences of y and of y^{-1} in $pp^{-1}y$ is odd; y and y^{-1} form the only pair of arrows of Γ^* for which this is true. Similarly z and z^{-1} are the only arrows which together occur an odd number of times in $pp^{-1}z$.

Consider the $\sim$-transitions. A transition using $(rr^{-1}ss^{-1}, ss^{-1}rr^{-1})$, $r,s \in P(\Gamma^*)$, does not change the number of occurrences of any arrow $a \in A \cup A^{-1}$; a transition involving $(rr^{-1}r,r)$, $r \in P(\Gamma^*)$, changes the number of occurrences of both a and a^{-1}, for any $a \in A \cup A^{-1}$, by the same number. So in all cases the parity of the sum of the number of occurrences of a and a^{-1} is preserved under $\sim$-transitions. Hence it follows that $y = z$ or $y = z^{-1}$.

Suppose $y = z^{-1}$. Then in pp^{-1}, if y^{-1} occurs n times, then y occurs $n + 1$ times in $pp^{-1}y$; whereas in $pp^{-1}z$, y^{-1} occurs $n + 1$ times and y occurs n times. Taking into account the above remarks on the effect of $\sim$-transitions it follows that any element obtained from $pp^{-1}y$ by $\sim$-transitions must contain $n + k$ occurrences of y^{-1} and $n + k + 1$ occurrences of y, for some integer k. Since $pp^{-1}z \sim^* pp^{-1}y$ we must therefore have $n + k = n + 1$ and $n + k + 1 = n$. This is impossible. So $y \neq z^{-1}$. Hence $y = z$; which completes the proof of the lemma.

The following lemma justifies the terminology already introduced.

LEMMA 3.6. *$\pi(\Gamma)$ is a primitive semigroup.*

Proof. Let $e = e^2 \in \pi(\Gamma)$, $f = f^2 \in \pi(\Gamma)$ and $f \leqslant e$, with $f \neq 0$. Then $fe = ef = f \neq 0$. Hence $o(f) = t(f) = o(e) = t(e)$. Since $\sigma^{\natural}$ is a morphism, there exist idempotents $g \in e$, $h \in f$, in $\kappa(\Gamma)$ and then $o(g) = t(g) = o(e)$ and $o(h) = t(h) = o(f)$. Hence, since $ef \neq 0$, it follows that $gh \neq 0$ and so $ghg = ghh$, since $\kappa(\Gamma)$ is inverse. Since gh is an idempotent, therefore $g\ \sigma\ h$, whence $e = f$.

LEMMA 3.7. *Let Γ be connected. Then $\pi(\Gamma)$ is completely 0-simple, i.e. a Brandt semigroup, and the structure group of $\pi(\Gamma)$ is a free group.*

Proof. The proof that $\pi(\Gamma)$ is a Brandt semigroup is straightforward and follows familiar lines. That each $\pi(\Gamma,c)$ is a free group is well-known.

4. Structure of $x(\Gamma)$

Let $F_{A\cup A^{-1}}$ denote the free semigroup on the set $A \cup A^{-1}$.

Consider the mapping

$$\alpha\colon F_{A\cup A^{-1}} \to P^0(\Gamma^*)$$

defined by

$$w = x_1x_2\ldots x_n \to x_1x_2\ldots x_n,$$

where each $x_i \in A \cup A^{-1}$ and where, on the left, $x_1x_2\ldots x_n$ is a word in $F_{A\cup A^{-1}}$, while, on the right, $x_1x_2\ldots x_n$ denotes the product in $P^0(\Gamma^*)$.

LEMMA 4.1. *α is a surjective morphism of $F_{A\cup A^{-1}}$ upon $P^0(\Gamma^*)$.*

Proof. Let $w_1, w_2 \in F_{A\cup A^{-1}}$ and let w_1w_2 denote the product in $F_{A\cup A^{-1}}$. Suppose $w_1 = x_1x_2\ldots x_n$ and $w_2 = y_1y_2\ldots y_m$, where each x_i, y_j belongs to $A \cup A^{-1}$. Then

$(w_1w_2)\alpha = 0$ if and only if one of

(i) $t(x_{i-1}) \neq o(x_i)$, $i = 2,\ldots,n$, i.e. if $w_1\alpha = 0$;

(ii) $t(y_{j-1}) \neq o(y_j)$, $j = 2,\ldots,m$, i.e. $w_2\alpha = 0$;

(iii) $t(x_n) \neq o(y_1)$, $w_1\alpha \neq 0$ and $w_2\alpha \neq 0$,
i.e. $t(w_1\alpha) \neq o(w_2\alpha)$.

Hence $(w_1w_2)\alpha = 0$ if and only if $(w_1\alpha)(w_2\alpha) = 0$.

If $(w_1w_2)\alpha \neq 0$, then clearly, $w_1\alpha \neq 0$, $w_2\alpha \neq 0$ and

$$(w_1w_2)\alpha = (w_1\alpha)(w_2\alpha).$$

Hence α is a morphism. It is clearly surjective as stated.

Consider the relation $\sim$ of section 3, but now defined on $F_{A\cup A^{-1}} = F$, say:

$$\sim_F = \sim = \{(yy^{-1}y,y) \mid y \in F\} \cup \{(yy^{-1}zz^{-1},zz^{-1}yy^{-1}) \mid y,z \in F\}.$$

It will cause no confusion to denote both relations by $\sim$. Recall that $F/{\sim^*}$ is the free inverse semigroup, which we denote by FI_A, on A.

LEMMA 4.2. *The diagram*

$$\begin{array}{ccc} F_{A\cup A^{-1}} & \xrightarrow{\quad\alpha\quad} & P^0(\Gamma^*) \\ \natural\big\downarrow & & \big\downarrow\natural \\ F/{\sim^*}=FI_A & \xrightarrow{\quad\alpha^*\quad} & P^0(\Gamma^*)/{\sim^*}=\kappa(\Gamma), \end{array}$$

where α^* is given by $w{\sim^*} \mapsto (w\alpha){\sim^*}$, commutes ($\alpha^*$ is the only map to make the diagram commute). Moreover, α^* is a surjective morphism.

Proof. First let us show that α^* is well-defined. Let $w \in F_{A\cup A^{-1}}$ and suppose that $w\alpha = 0$ and let $w' \sim^* w$. Then w' is obtained from w by a sequence of $\sim$-transitions. A transition induced by a pair $(yy^{-1}y,y)$ clearly preserves the property of being mapped to 0 by α. A transition induced by a pair $(yy^{-1}zz^{-1},zz^{-1}yy^{-1})$ is of the form

$$w = w_1yy^{-1}zz^{-1}w_2 \to w_1zz^{-1}yy^{-1}w_2 = w''.$$

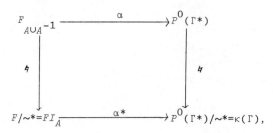

If $w_1\alpha = 0$ or $w_2\alpha = 0$, then $w\alpha = w''\alpha = 0$. If $(yy^{-1}zz^{-1})\alpha = 0$, then in $P^0(\Gamma)$, $o(yy^{-1}) = t(yy^{-1}) \neq o(zz^{-1}) = t(zz^{-1})$. Hence $(zz^{-1}yy^{-1})\alpha = 0$. If $w_1\alpha \neq 0$ and $(yy^{-1}zz^{-1})\alpha \neq 0$, then $o(y) = o(yy^{-1}) = t(yy^{-1}) = o(z) = o(zz^{-1})$. Hence if $(w_1yy^{-1}zz^{-1})\alpha = 0$, then $t(w_1\alpha) \neq o(yy^{-1}zz^{-1}) = o(zz^{-1}yy^{-1})$ and so also $(w_1zz^{-1}yy^{-1})\alpha = 0$. Similarly, if $(yy^{-1}zz^{-1})\alpha \neq 0$, $w_2\alpha \neq 0$ and $(yy^{-1}zz^{-1}w_2)\alpha = 0$, then $(zz^{-1}yy^{-1}w_2)\alpha = 0$. Hence, in all cases, $w\alpha = 0$ implies $w''\alpha = 0$.

Thus, if $w\alpha = 0$, then $w{\sim}^* \mapsto (w\alpha){\sim}^*$ is well-defined.

If $w\alpha \neq 0$, then $w' \sim^* w$ in $F_{A \cup A'}$ if and only if $w'\alpha \sim^* w\alpha$ in $P^0(\Gamma)$. Hence α^* is well-defined. It is the unique mapping that makes the diagram commute, for its definition just asserts this commutativity. Since the left-most vertical mapping in the diagram is onto, it follows that α^* is a morphism. Clearly α^* is surjective.

THEOREM 4.3. *$\kappa(\Gamma)$ is a morphic image of $FI_{A(\Gamma)}$. Indeed*

$$\kappa(\Gamma) \simeq FI_A/0(\alpha^{*-1}),$$

a Rees ideal quotient of $FI_A = FI_{A(\Gamma)}$.

Proof. The preceding lemma gives $\kappa(\Gamma)$ as the image under α^* of FI_A. Moreover $0(\alpha^{*-1})$ is an ideal of FI_A and, if $(w_1{\sim}^*)\alpha^* = (w_1\alpha){\sim}^* = (w_2\alpha){\sim}^* = (w_2{\sim}^*)\alpha^* \neq 0$, then, as shown in the proof of lemma 4.2, $w_1\alpha = w_2\alpha$ and $w_1{\sim}^* = w_2{\sim}^*$, i.e. α^* is one-to-one outside $0(\alpha^{*-1})$.

THEOREM 4.4. *$\kappa(\Gamma)$ is (combinatorial or) H-degenerate.*

Proof. Since FI_A is H-degenerate, and H-degeneracy is preserved under the taking of Rees-quotients, this result is an immediate corollary to theorem 4.3.

COROLLARY 4.5. *If $x \in \kappa(\Gamma)$, then x is idempotent if and only if $x = x^{-1}$.*

Proof. This is true for any H-degenerate semigroup. For, if $x = x^2$, then $x = x^{-1}$ in any inverse semigroup. And conversely, if $x = x^{-1}$, then $xx^{-1} = x^{-1}x$ and so, since $\kappa(\Gamma)$ is H-degenerate, $x = x^2 \in H_{xx^{-1}}$, the sole element of its H-class.

THEOREM 4.6. *Let* Γ_i, $i \in I$, *be the connected components of* Γ. *Then* $\kappa(\Gamma)$ *is the 0-direct union of the* $\kappa(\Gamma_i)$, $i \in I$.

Proof. If $p \in P^0(\Gamma^*)$, then there exists Γ_i such that $p \in P^0(\Gamma_i^*)$, and if $i \neq j$, then $P^0(\Gamma_i^*) \cap P^0(\Gamma_j^*) = \{0\}$. Moreover, if two elements of $P^0(\Gamma^*)$ are $\sim$*-equivalent, then examining $\sim$ shows that they both belong to some $P^0(\Gamma_i^*)$. Hence $\kappa(\Gamma_i) \cap \kappa(\Gamma_j) = \{0\}$, if $i \neq j$, and $\kappa(\Gamma) = \cup\{\kappa(\Gamma_i) | i \in I\}$.

Furthermore, if $p \in P^0(\Gamma_i^*)$ and $q \in P^0(\Gamma_j^*)$ with $i \neq j$, then $pq = 0$. Hence $\kappa(\Gamma_i).\kappa(\Gamma_j) = 0$, when $i \neq j$; which completes the proof of the theorem.

THEOREM 4.7. *Each* $\kappa(\Gamma, v)$, *for* $v \in V^0(\Gamma)$ *is (isomorphic to) a closed inverse subsemigroup of* FI_A.

Proof. By theorem 4.3, $\kappa(\Gamma, v)$ is effectively a subsemigroup of FI_A; let us regard it as such.

Let $x \in \kappa(\Gamma, v)$ and $x \leqslant y$, so that $x = ey$, for some idempotent e of $\kappa(\Gamma)$. Let $p, q, w \in P(\Gamma)$ such that $p \in x$, $q \in y$ and $ww^{-1} \in e$. Thus $p \sim^* ww^{-1}q$. Hence $o(p) = o(ww^{-1}q) = o(ww^{-1}) = t(ww^{-1}) = o(q)$ and $t(p) = t(q)$. Hence $q\sim^* = y \in \kappa(\Gamma, v)$.

5. Conclusion

All that has been done in this paper is to present various semigroups and inverse semigroups associated with any graph. It is hoped that this provides the right scenario to use, conversely, to investigate some of the properties of inverse semigroups and in particular of the various free objects in the category of inverse semigroups.

References

1971 S. Mac Lane, *Categories for the working mathematician.* Graduate Texts in Mathematics 5, Springer-Verlag, New York, 1971.

1973 W.D. Munn, Free inverse semigroups. *Proc. London Math. Soc.* 29 (1974), 385-404.

1979 H.H. Teh and S.C. Shee, *Colour graphs and inverse semigroups*. Research report no.83, Lee Kong Chian Inst. of Math. and Computer Sci., Nanyang University, 1979, pp.1-17.

THE FORMAL STRUCTURE OF OBSERVATIONAL PROCEDURES

P. D. Finch

Department of Mathematics, Monash University, Clayton, Victoria, Australia, 3168.

Abstract

Observational procedures are characterised in terms of certain subsets of the Cartesian product $P_* \times R_*$ of free semigroups generated by a set of pointers P and a set of readings R. Experiments, designs, objects, systems and states are defined in terms of pointer-readings. Temporal development in a non-relativistic setting is defined by a one-parameter semigroup of transformations of state-space.

It is shown that the structure of what can be known by observation can be classified in two ways according as the results of later observations do or do not depend on whether we make earlier ones. The latter is the so-called classical case, the former is referred to as non-classical. The classical case reduces to the population model of classical statistics whereas the structure of the non-classical case is that of a sequential computing machine.

Some formal analogies between the structure of observational procedures and the standard methods of quantum mechanics are discussed very briefly.

1. Introduction

We start with two things: non-empty sets P and R whose elements are called unary pointers and unary readings. These primitive terms are undefined but their use is meant to suggest the obvious interpretation. A unary pointer is a device for observing something whereas a unary reading is what is displayed by such a device when it is used in that way. Strictly speaking a unary pointer is an equivalence class of identical devices and a unary reading is the common reading each of them would exhibit in any given case. In other words a unary pointer is a piece of software, it refers to the function performed by a device rather than to a particular piece of hardware. However we do not exclude the possibility that some pieces of hardware are unique. The

ISBN 0-12-319450-4

formal development is consistent with the view that each observational device is unique in the sense that no other one would exhibit exactly the same unary reading in any given instance of its use.

Multiple pointers are finite sequences of unary pointers, they comprise P_* the free semigroup generated by P. The length of the word defining p in P_* is its arity and $p > q$ means p extends q. The intended interpretation is that a multiple pointer denotes consecutive uses of individual devices. The uses in question are those indicated by the individual unary pointers in its defining expression $p_1p_2...p_n$. The order of those uses is the left-to-right ordering of that expression. In like manner R generates R_*, the set of multiple readings. An n-ary reading $r_1r_2...r_n$ is the set of readings associated with an n-ary pointer $p_1p_2...p_n$. The unary reading r_k is then the reading on the unary pointer p_k when the consecutive usages of the unary pointers $p_1, p_2, ..., p_n$ are those indicated by the expression $p_1p_2...p_n$. We say that $r_1r_2...r_n$ is the reading on the pointer $p_1p_2...p_n$. The set of n-ary pointers is P_n, that of n-ary readings is R_n. An element (p,r) of $P_n \times R_n$ is called an n-ary pointer-reading. The set of all pointer-readings is denoted by PR. The projections $PR \to P_*$, $PR \to R_*$ are denoted by Π_P and Π_R, $\Pi_P(p,r) = p$ and $\Pi_R(p,r) = r$. The intended interpretation is the obvious one: a pointer-reading is a multiple pointer displaying a multiple reading.

In practice we are not involved with all pointer-readings but only some of them. This suggests that the formal counterpart to an experiment should be a designated subset of PR. But an experiment has to be planned and its design includes not only a list of the observational devices in question but also instructions on the way they are to be used. A design schema $\mathcal{D}$ is a non-empty subset of P_* with the property

$$p \in \mathcal{D} \ \&\ p > q \Rightarrow q \in \mathcal{D}. \tag{1.1}$$

It names the devices under consideration and specifies the way they are to be used. We introduce (1.1) to conform to our interpretation. If an n-ary pointer $p_1p_2...p_n$ corresponds to consecutive observations by unary pointers, then the successive stages of that observational process are also part of the experiment in question. In other words, the design schema should contain each of the pointers $p_1, p_1p_2, ...,$ $p_1p_2...p_{n-1}$ whenever it contains the pointer $p_1p_2...p_n$.

An experiment schema E is a subset of PR with $\Pi_P E$ a design schema and, in addition,

$$(p,r) \in E \ \& \ (p,s) \in E \Rightarrow r = s \tag{1.2}$$

and

$$(p,r) \in E,\ (q,s) \in E \ \& \ p > q \Rightarrow r > s \tag{1.3}$$

The motivation is that an experiment schema is a design schema together with the readings displayed by the pointers in question. The informal requirement that a particular observational device does not exhibit more than one reading at any one use of it leads to (1.2). Similarly (1.3) means that later observations do not change the readings obtained from earlier ones. The two implications taken together yield

$$(p,r) \in E,\ (q,s) \in E \ \& \ p \geqslant q \Rightarrow r \geqslant s \tag{1.4}$$

It follows from (1.2) that E is a function mapping $\mathcal{D}$ into R_*. That function is denoted by α and written on the right so that, for p in $\mathcal{D}$, $p\alpha$ is the unique reading in R_* for which $(p,p\alpha)$ is in E. We say that α is the reading function of E, it assigns readings to the pointers in the design schema $\mathcal{D}$, from (1.4) it preserves arity and extension. Conversely if $\mathcal{D}$ is a design schema and $\alpha\colon \mathcal{D} \to R_*$ has domain $\mathcal{D}$ and preserves both arity and extension, then $E = \cup_{\mathcal{D}}(p,p\alpha)$ is an experiment schema with reading function α and design schema $\mathcal{D}$. Thus the concept of an experiment schema is equivalent to that of an ordered pair $(\mathcal{D},\alpha)$ consisting of a design schema $\mathcal{D}$ and a reading function α, viz. a mapping of $\mathcal{D}$ into R_* which preserves arity and extension. It is often convenient to specify an experiment schema in that way.

Let $\mathcal{D}$ be a design schema and let p be a pointer. If $\mathcal{D}_{|p} = \{q\colon pq \in \mathcal{D}\}$ is not empty it is a design schema. It is said to be the conditional design schema determined by p. If p is not in $\mathcal{D}$ then $\mathcal{D}_{|p}$ is empty. A p in $\mathcal{D}$ for which $\mathcal{D}_{|p}$ is empty is said to be terminal in $\mathcal{D}$. Let $E = (\mathcal{D},\alpha)$ be an experiment schema. For a p in $\mathcal{D}$ which is not terminal, $\alpha_{|p}\colon \mathcal{D}_{|p} \to R_*$ is defined by decreeing that, for q in $\mathcal{D}_{|p}$, $q\alpha_{|p}$ is the unique reading in R_* such that

$$(pq)\alpha = (p\alpha)(q\alpha_{|p}). \tag{1.5}$$

The reading $q\alpha_{|p}$ is said to be the conditional reading on q generated by p. We sometimes refer to it as a readout. It is obvious that $\alpha_{|p}$ is a reading function for $\mathcal{D}_{|p}$; $E_{|p} = (\mathcal{D}_{|p}, \alpha_{|p})$ is said to be the conditional experiment schema generated from $E = (\mathcal{D}, \alpha)$ by p.

A reading function is a function on the pointers in a design schema. It is not necessarily a function on the pointers which occur as parts of the expressions defining the pointers in the design schema, even though some of those parts are pointers in that design schema. For example, q in (1.5) may or may not belong to $\mathcal{D}$ and it may or may not occur in another expression such as mq in $\mathcal{D}$. If mq is in $\mathcal{D}$, then $q\alpha_{|m}$ is also defined and if q is in $\mathcal{D}$, then $q\alpha$ is a well-defined reading. But there is no conflict with the requirement that each observational device exhibit only one reading at any one use of it because the symbol q in the expressions q, pq and mq denotes different uses of that pointer.

2. Objects and their unfolding

In scientific discourse we often talk about experiments in terms of the effects produced by the application of observational procedures to physical objects. The use of such imagery is helpful because it enables us to refer to complicated sets of operations in a compact but meaningful way. In the formal system objects are constructs of pointer-readings. We do not assert they are 'things' out there in the world existing independently of what we observe about them. On the other hand we are not denying there could be some meaning to such an assertion. Perhaps our everyday way of talking about physical objects is a poetic paraphrase of the constructive procedure described below.

Let $E = (\mathcal{D}, \alpha)$ be an experiment schema and let p be a pointer. The set

$$Ob\{E,p\} = \{(q, q\alpha): q \in \mathcal{D} \ \& \ q \geqslant p\} \qquad (2.1)$$

is said to be the object of E determined by p. The empty set is the null object, $Ob\{E,p\}$ is not empty if and only if p is in $\mathcal{D}$. The correspondence $p \leftrightarrow Ob(E,p)$ is one-one on $\mathcal{D}$. The arity of a non-null object is the arity of the pointer which determines it. Two objects of E are separated when their set intersection is empty. Objects which are not separated are said to be connected. The following result is immediate:

PROPOSITION 2.1. *Non-null objects* $Ob\{E,p\}$ *and* $Ob\{E,q\}$ *are connected if and only if either* $p \geqslant q$ *or* $q \geqslant p$.

The non-null object $Ob\{E,p\}$ is said to succeed $Ob\{E,q\}$ when $p \geqslant q$ in which case $Ob\{E,p\} \subseteq Ob\{E,q\}$. Suppose $Ob\{E,p\}$ has arity $n > 1$. For each positive integer $m < n$ there is a unique object $Ob\{E,q\}$ of arity m which is succeeded by $Ob\{E,p\}$. Two objects of different arities are separated unless one succeeds the other and any two distinct objects with the same arity are separated. We have then

PROPOSITION 2.2. *An experiment schema is the set union of the separated unary objects determined by the unary pointers in its design schema.*

A non-null object is not something given *a priori* which is subsequently observed. It is something constructed out of the observational process and contains within itself all the objects it might become during a subsequent unfolding of that process. The pointer-reading $(p,p\alpha)$ describes how the object $Ob\{E,p\}$ came to be. But that object is potentially any one of the objects which succeed it in the sense that if pq is in $\mathcal{D}$, then the pointer pq determines the new object $Ob\{E,pq\}$. Thus each pointer q in P_* acts as an operator or object shift function which sends $Ob\{E,p\}$ into $Ob\{E,pq\}$. We exhibit these shifts by writing the pointer to the right of the object on which it operates. If $\mathcal{O} = \mathcal{O}(E)$ is the set of objects of E, the object shift function $q:\mathcal{O} \to \mathcal{O}$ is given by

$$Ob\{E,p\}q = Ob\{E,pq\}. \tag{2.2}$$

We say that q is applicable to $Ob\{E,p\}$ when $Ob\{E,pq\}$ is non-null, viz. when p, pq are in $\mathcal{D}$ and q is in $\mathcal{D}_{|p}$.

The observational process involves more than the unfolding of an object into other objects. At each stage of that process there is a corresponding readout. In other words a pointer q can also act as an object readout function in accordance with the expression

$$\forall\ p \in \mathcal{D}\ \&\ \forall\ q \in \mathcal{D}_{|p}\ :\ qOb\{E,p\} = q\alpha_{|p}. \tag{2.3}$$

In (2.3) q is written on the left of the object to avoid confusion with its role in (2.2). Thus we regard the pointer q in two different ways. Firstly as an object shift operator which maps objects to objects, and we then put it to the right of the operand. Secondly as an object readout function, we then write it to the left of its argument.

Observation is an operation on objects which yields both readouts and new objects. An object is always changed by observation. It is never the same object once it has been observed. Nevertheless an object may be effectively unchanged by observation inasmuch as readouts from future observations may not depend on past and present ones. Suppose that b is applicable to $Ob\{E,p\}$ and that c is applicable to both $Ob\{E,p\}$ and $Ob\{E,pb\}$. We say that $Ob\{E,p\}$ is effectively unchanged by b relative to c when

$$c\alpha_{|pb} = c\alpha_{|p}. \tag{2.4}$$

It is sometimes convenient to regard E itself as one of its own objects. We assign it zero arity and call it the nullary object of E. It should not be confused with the null object which is the empty set. When E is included in $O(E)$ we define Eq and qE by decreeing that

$$\forall\ q \in \mathcal{D}\colon\ Eq = Ob\{E,q\}\ \&\ qE = q\alpha \tag{2.5}$$

In this context we sometimes write $E = Ob\{E,p_0\}$ by introducing a unique nullary pointer p_0. This is equivalent to adjoining an identity element to the semigroup P_*, $p_0p_0 = p_0$ and $pp_0 = p = p_0p$ for all p in P_*. Equations (2.5) are then the special cases of (2.2) and (2.3) with $p = p_0$. In the same way, taking $p = p_0$ in (2.4) and changing the notation slightly, we have

PROPOSITION 2.3. *If $E = (\mathcal{D},\alpha)$ is an experiment schema and q, pq are both in $\mathcal{D}$, then E is effectively unchanged by p relative to q when*

$$q\alpha_{|p} = q\alpha, \tag{2.6}$$

that is to say when

$$(pq)\alpha = (p\alpha)(q\alpha)$$

In particular, when $\mathcal{D}$ is a subsemigroup of P_*, E is effectively unchanged by any p relative to any q in $\mathcal{D}$ if and only if α is a semigroup morphism of $\mathcal{D}$ into R_*.

3. Systems and states

Scientific dialogue is often simplified by talking as if observation involved changes of state rather than the unfolding of new objects. We shall identify states with conditional reading functions. Moreover we gain access to a more familiar vocabulary by introducing the concept of a 'system'. A system in an initial state will turn out to be effectively the same thing as an experiment schema. The important thing about a 'system' is that it retains its identity throughout any one of the sequences of observation defined in a design schema and yet can still be said to exhibit changes of state and changes of configuration. We separate the notions of system and state by defining the former in terms of certain 'configurations'.

Let $\mathcal{D}$ be a design schema. We take configurations to be the conditional design schemas $\mathcal{D}_{|p}$ with p in $p_0 \cup \mathcal{D}$. The initial configuration is defined to be $\mathcal{D}$ itself. At the informal level a system is something which retains its identity from the standpoint of the ordered pairs $(p,\mathcal{D}_{|p})$ with p in $p_0 \cup \mathcal{D}$. The pointer p names a sequence of unary observations and $\mathcal{D}_{|p}$ names the configuration brought about by making them. It is a short step from this to identify the system in question with the set $\{(p,\mathcal{D}_{|p})\colon p \in p_0 \cup \mathcal{D}\}$ and thereby *define* the system determined by $\mathcal{D}$ to be the function mapping p in $p_0 \cup \mathcal{D}$ to the conditional design schema $\mathcal{D}_{|p}$. This is the definition we adopt. We use the same symbol to denote a design schema and the system it determines. In other words we regard the symbol $\mathcal{D}$ in two ways. As a set of pointers it is a design schema, as a function it is a system. The system is the function defined on $p_0 \cup \mathcal{D}$ by the expression

$$\forall\, p \in p_0 \cup \mathcal{D}\colon\ \mathcal{D}(p) = \mathcal{D}_{|p} \tag{3.1}$$

The conditional design schemas $\mathcal{D}_{|p}$ are said to be the configurations of the system $\mathcal{D}$, $\mathcal{D}_{|p}$ is the configuration determined by p. In particular, $\mathcal{D}$ itself is the configuration determined by p_0, it is said to be the initial configuration of the system. The set of all configurations of the system $\mathcal{D}$ is denoted by $conf(\mathcal{D})$. Each configuration is a design schema and determines a corresponding system.

Any reading function for the design schema $\mathcal{D}$ is said to be a possible initial state of the associated system. The set of all initial states is denoted by $ISt(\mathcal{D})$. It is called the initial state schema of the system $\mathcal{D}$. The set

$$St(\mathcal{D}) = \cup_{\mathcal{D}' \in Conf(\mathcal{D})} ISt(\mathcal{D}')$$

is called the state schema of the system $\mathcal{D}$. Its elements are referred to as states of the system $\mathcal{D}$. The state schema $St(\mathcal{D})$ is the set of functions s for which there is p in $\mathcal{D}$ and α in $ISt(\mathcal{D})$ with $s = \alpha_{|p}$, in particular s has domain $\mathcal{D}_{|p}$. If s is a state of $\mathcal{D}$ there is exactly one configuration $\mathcal{D}'$ of $\mathcal{D}$ for which s in an initial state of $\mathcal{D}'$, viz. the domain of s. But there may be more than one pointer p in $\mathcal{D}$ with $\mathcal{D}_{|p} = \mathcal{D}'$ and $\alpha_{|p} = s$.

Let s belong to $St(\mathcal{D})$. By 'the system $\mathcal{D}$ in the state s' we mean the ordered triple $(\mathcal{D},\mathcal{D}',s)$ where $\mathcal{D}'$ is the configuration of $\mathcal{D}$ for which s is an initial state, viz. the domain of s. In particular 'the system $\mathcal{D}$ in the initial state α' is the triple $(\mathcal{D},\mathcal{D},\alpha)$ or, more simply, the ordered pair $(\mathcal{D},\alpha)$ if we take it as understood that α is then an initial state of $\mathcal{D}$. In other words, as we said before, a system in an initial state is effectively the same thing as an experiment schema or, equivalently, a nullary object.

It is sometimes convenient to transfer attention from the system $\mathcal{D}$ in the state $\alpha_{|p}$, viz. the triple $(\mathcal{D},\mathcal{D}_{|p},\alpha_{|p})$, to the system $\mathcal{D}_{|p}$ in the initial state $\alpha_{|p}$, viz. $(\mathcal{D}_{|p},\alpha_{|p})$. It should be noted, however, that $\alpha_{|p}$ pertains to future and not past observations. Specifying $\mathcal{D}_{|p}$ and $\alpha_{|p}$ does not, in itself, name the pointer p, nor does it record the reading $p\alpha$. An initial state tells us what happens in each of the futures under consideration. It does not tell us how we got into that state and it does not record what we observed getting there. A non-null object $Ob\{E,p\}$ is essentially an ordered quadruple $(p,p\alpha,\mathcal{D}_{|p},\alpha_{|p})$ whereas the associated system in the initial stage $\alpha_{|p}$ is the pair $(\mathcal{D}_{|p},\alpha_{|p})$. An object is a system in an initial state together with the pointer specifying how it arose and the data comprising the past reading $p\alpha$. This distinction is not immediately obvious when E, regarded as a nullary object, is taken as the system in a given initial state because p is then p_0 and there is no past to be taken into account. The system is, as it were, a newly formed embryo, as yet without a past but with a fully developed potential for what is still to come.

The system-state terminology leads to corresponding changes in the way we talk about the object shifts and readouts associated with observation. If $(\mathcal{D},\alpha)$ is the system $\mathcal{D}$ in the initial state α and p is in $p_0 \cup \mathcal{D}$, then each pointer q in $\mathcal{D}_{|p}$ acts both as a state transition operator and a state readout function in accordance with expressions which are straightforward analogues of (2.2) and (2.3). Thus (2.2) takes the form

$$\forall\ p \in p_0 \cup \mathcal{D}\ \&\ \forall\ q \in \mathcal{D}_{|p}\colon\ \alpha_{|p}q = \alpha_{|pq} \tag{3.2}$$

and exhibits that state transition $\alpha_{|p} \to \alpha_{|pq}$ brought about by observing $\mathcal{D}$ in the state $\alpha_{|p}$ by means of the pointer q. In like manner, (2.3) becomes

$$\forall\ p \in p_0 \cup \mathcal{D}\ \&\ \forall\ q \in \mathcal{D}\colon\ q(\alpha_{|p}) = (q)\alpha_{|p}. \tag{3.3}$$

The brackets emphasise that the left-hand side is the result of operating on a state with a pointer whereas the right-hand side is the result of operating on a pointer with a state. Both operations yield a reading.

Let s belong to $St(\mathcal{D})$. There is p in $p_0 \cup \mathcal{D}$ and α in $ISt(\mathcal{D})$ with $s = \alpha_{|p}$. If the system $\mathcal{D}$ is observed by q in $\mathcal{D}_{|p}$ when it is in the state s there is a state change $s \to sq$ given by

$$\forall\ q \in \mathcal{D}_{|p}\colon\ sq = \alpha_{|pq}. \tag{3.4}$$

The associated readout is qs given by

$$\forall\ q \in \mathcal{D}_{|p}\colon\ qs = q\alpha_{|p}. \tag{3.5}$$

These definitions do not depend on the p in $\mathcal{D}$ and the α in $ISt(\mathcal{D})$ for which $s = \alpha_{|p}$. To see this note that for α,β in $ISt(\mathcal{D})$

$$\alpha_{|p'} = \beta_{|p''} \Rightarrow \mathcal{D}_{|p'} = \mathcal{D}_{|p''}$$

and, for all q in $\mathcal{D}_{|p'} = \mathcal{D}_{|p''}$,

$$\alpha_{|p'} = \beta_{|p''} \Rightarrow q\alpha_{|p'} = q\beta_{|p''}\ \&\ \alpha_{|p'q} = \beta_{|p''q}\ .$$

That the antecedent in this implication implies the first equality in the consequent is obvious. The second equality follows from the first one and the fact that, for any x in $\mathcal{D}_{|p'q} = \mathcal{D}_{|p''q}$, one has

$$(q\alpha_{|p'})(x\alpha_{|p'q}) = qx\alpha_{|p'} = qx\beta_{|p''} = (q\beta_{|p''})(x\beta_{|p''q}).$$

Although states are functions mapping multiple pointers to multiple readings it is only necessary in practice to specify the *unary* states, viz. their restrictions to the unary pointers in their respective domains. This follows from the multiplicative law

$$q_1q_2\cdots q_ns = (q_1s)(q_2sq_1)(q_3sq_1q_2)\cdots(q_nsq_1q_2\cdots q_{n-1}) \tag{3.6}$$

Here $q_msq_1q_2\cdots q_{m-1}$ denotes the unary readout from the unary pointer q_m when the system is in the state $sq_1q_2\cdots q_{m-1}$. When we know *all* the unary states of a system we can reconstruct its general states from (3.6). If s is a state its unary restriction is denoted by sU.

4. Simple systems and sequential machines

An important type of system arises when the design schema $\mathcal{D} = D_*$, the free semigroup generated by a subset D of P. Each unary pointer in D is called a dimension of the system, D is said to be its dimensional support and D_* is said to be a simple system. When $\mathcal{D} = D_*$ we have

$$\mathcal{D}_{|p} = \mathcal{D} = D_*, \ \forall\ p \in p_0 \cup \mathcal{D}$$

and so D_* has only one configuration, viz. the initial one. The states of D_* are all initial states, they are the functions $s\colon D_* \to R_*$ mapping D_* into R_* with preservation of arity and extension. The following result is easily proved.

PROPOSITION 4.1. *Let $\mathcal{D}$ be a system, let D be the set of unary pointers in $\mathcal{D}$ and let D' be the set of unary pointers which appear as components of the pointers in $\mathcal{D}$. Then*

(i) $D \subseteq D'$

(ii) $\mathcal{D} \subset D'_*$

(iii) any system is a sybsystem of a simple system

(iv) $pq \in \mathcal{D}\ \&\ \mathcal{D}_{|p} = \mathcal{D} \Rightarrow q \in \mathcal{D}$

(v) $\mathcal{D}_{|p} = \mathcal{D},\ \forall\ p \in \mathcal{D} \Rightarrow D' = D$

(vi) $\mathcal{D}$ is simple if and only if $\mathcal{D}_{|p} = \mathcal{D}$ for each p in $\mathcal{D}$ and then $\mathcal{D} = D_*$.

Let $\mathcal{D}$ be a system and let α belong to $ISt(\mathcal{D})$. We say that α is classical at p in $\mathcal{D}$ when $\alpha_{|p} = \alpha$. An initial state α is classical at p if and only if

$$\mathcal{D}_{|p} = \mathcal{D} \quad \& \quad \forall\, q \in \mathcal{D}:\ q\alpha_{|p} = q\alpha,$$

or equivalently

$$\mathcal{D}_{|p} = \mathcal{D} \quad \& \quad \forall\, q \in \mathcal{D}:\ pq\alpha = (p\alpha)(q\alpha).$$

An α in $ISt(\mathcal{D})$ is classical when it is classical at each p in $\mathcal{D}$. If a system admits classical states we must have $\mathcal{D}_{|p} = \mathcal{D}$ for all p in $\mathcal{D}$ and so, by proposition (4.1), the system must be simple. But simple systems also admit non-classical states. From proposition (2.3) we obtain

PROPOSITION 4.2. *A state* α *of a simple system* D_* *is classical if and only if it is a semigroup morphism of* D_* *into* R_*.

The unary states of a simple system D_* are functions mapping D into R. To specify all the states it is sufficient to name the unary states. If we know the $q\alpha_{|p}$ for all p in D_* and all q in D, then we can compute the $q\alpha_{|p}$ for p and q both in D_* by means of (3.6). When α is a classical state (3.6) gives

$$\forall\, q_1, q_2, \ldots, q_n \in D:\ q_1 q_2 \ldots q_n\alpha = (q_1\alpha)(q_2\alpha)\ldots(q_n\alpha). \tag{4.1}$$

Thus a classical state is completely determined by its *own* unary restriction and does not depend on the general involvement with the unary restrictions of other states exhibited in (3.6). It follows that a simple system D_* in a classical state α can be thought of as an ordered pair $(D, \alpha U)$, where $\alpha U = \alpha|D$, the restriction of α to D, is the unary state determined by α. In other words, the concept of a simple system in a classical state reduces to that of a *population*, viz. an indexed set of unary readings $\{d\alpha:\ d \in D\}$.

It is only a short step from the population concept to thinking of the unary readings $d\alpha$ as things which characterise the system D_* in the state α independently of whether we observe it or not. But that step involves a failure to distinguish between that which does not depend on *how* we observe it and that which does not depend on *whether or not* we observe it. Strictly speaking we are still involved with sequential observation even in the case of a simple system in a classical state. Nevertheless it is often convenient to talk as if there is something out there which persists whether or not we observe it. But this is as much a matter of literary style as a commitment to an existential metaphysic.

A simple system in a non-classical state cannot be pictured as a population because what we observe depends on how we go about observing it. A more appropriate picture is that of a sequential machine in a given state. For our purposes such a machine M consists of a set of 'states' Σ, a set of inputs D, a set of outputs R, a state transition function $\gamma: \Sigma \times D \to \Sigma$ and an output function $\omega: \Sigma \times D \to R$. On receiving the input d when in the state σ, M emits the output $\omega(\sigma,d)$ and moves into the state $\gamma(\sigma,d)$. Identifying Σ with St, input words with pointers and output words with readings, a simple system in a non-classical state is equivalent to a machine with distinguishable states which are accessible from the initial state in question. Identification of σ in Σ with s in St shows that $\gamma(\sigma,d)$ corresponds to the new state sd whereas $\omega(\sigma,d)$ corresponds to the readout ds.

The study of simple systems is effectively that of sequential machines. We adopt the system-state terminology because it corresponds more closely to the type of practical problem we wish to study. Since any system is a subsystem of a simple system, the general case can be pictured as a sequential machine in the operation of which we are restricted to using only certain designated input words. For ease in exposition we shall, in the main, consider only simple systems.

5. Incomplete specification and pragmatic indeterminism

In this section D_* is a simple system with finite dimensional support $D = \{d_1, d_2, \ldots, d_K\}$. If D_* is in the state α, then all possible future states and readouts are determined through the equations

$$\forall\ pq \in D_*:\ \alpha p = \alpha_{|p}\ \&\ q(\alpha p) = q\alpha_{|p}.$$

Specification of the initial state α is a complete blueprint for the results of all the possible futures then under consideration. In other words a simple system in a state is a deterministic entity. In practice, however, things are not as simple as this. Though the formal framework is conceptually determinate it is, in practice, pragmatically indeterminate. Conceptual determinism means we can name the set of possible states whereas pragmatic indeterminism means that we cannot point to the particular element of that set which is the *actual* state.

In practice the initial state is incompletely specified because certain states in St are indistinguishable from each other at the practical level of observational enquiry. There is nothing mysterious in this sort of indeterminism. Though the initial state α is a complete blueprint for the results of all possible observations we do not usually perform all of them and so we cannot give an unambiguous name to that state. For instance, in some situations we are restricted by the practical context to the use of a single multiple pointer q in D_*. Even though we observe $r = q\alpha$ we cannot, in general distinguish between the actual state α and the other states β with $q\beta = r$. We can only name the actual state ambiguously. We can say of it only that it is one of the states for which the reading given by q is the one actually obtained. Pragmatic indeterminism is therefore the norm. It is its absence rather than its presence which is the exception.

Indeterminism is absent when the actual state is known to be classical and the observational pointer is $d_1 d_2 \ldots d_K$. For if the observed reading is then $r_1 r_2 \ldots r_K$ we have $r_k = d_k \alpha$ and α, being a semigroup morphism of D_* into R_*, is determined by its restriction to D. But even with classical states there are commonly occurring situations involving pragmatic indeterminism. We may be using a multiple pointer $q = q_1 q_2 \ldots q_n$ which involves only some of the dimensions in D. This is the sort of restriction studied in classical statistics. The individual terms of the n-ary reading $q\alpha$ are then called the sample values. But even when $q = d_1 d_2 \ldots d_K$ we might be restricted by choice, economy or force of circumstance to working, not with the whole record $q\alpha$ but only a summarising description of it. General discussion of pragmatic indeterminism involves an ordered pair (I, ρ) consisting of a set of states I and a binary relation ρ on I. Both I and ρ form part of the statement of the problem being studied, I is the set of initial states then under consideration and $s' \rho s''$ means that s', s'' in I are indistinguishable at the level of the enquiry in question. For ease of exposition we shall take ρ to be an equivalence. Indistinguishable states will be said to be similar.

A typical instance of similarity arises in the following way. Let D_* be a simple system and let Q be a designated subset of D_*. States α', α'' in $St(D_*)$ are said to be Q-similar when $q\alpha' = q\alpha''$ for all q in Q. When we are dealing with Q-similar states it is their common behaviour on Q which constitutes the similarity between them whereas it is their possibly different behaviour outside Q which gives rise to our talking about differences between them and the variability exhibited by otherwise similar things. If α',α'' are Q-similar we do not distinguish between them by observation with q in Q. They appear the same because observations of (D_*,α') and (D_*,α'') by q lead to the same readout. But if we subsequently observe them with a pointer m for which qm is not in Q, then we will in general get different readouts. This is the variability exhibited by otherwise similar things and its discovery has sometimes been a source of puzzlement in the physical sciences because of the practice of referring to vectors of unary readouts as 'states'. The difficulty arises when one is involved with D-similarity between states or, as we shall call it local similarity. We shall now discuss it in some detail.

In some areas of science a simple system D_* in the state α is viewed as an entity which is characterised by the 'values' of its various dimensions $d_1,d_2,\ldots,d_K$ and these are taken to be the corresponding unary readings $d_1\alpha,d_2\alpha,\ldots,d_K\alpha$. The entity itself is then represented by the vector $(d_1\alpha,d_2\alpha,\ldots,d_K\alpha)$. It is sometimes said that the entity is in the 'state' named by that vector or even that the vector is the 'state. This involves a use of the term 'state' which is different from the one adopted here. In our terminology the vector representation is the unary state of the system and its misuse in characterising that system involves the misapplication of classical concepts to non-classical contexts. The vector representation of a 'state' seems to be suggested by two things:

(A) for each $k = 1,2,\ldots,K$ it is the case that if the 'value' of the dimension d_k were to be determined when the system was in the state α, then that value would turn out to be $d_k\alpha$, and

(B) these values 'characterise' the entity under consideration inasmuch as, taken together, they fully describe it because, by assumption, there are no other dimensions to be taken into account.

But in (A) we are dealing with a parallel set of potential observations involving a disjunction of unary pointers: either we observe with d_1 or with d_2, and so on. In (B), on the other hand, we are dealing with a full characterisation of the entity in question and this involves one with sequential rather than parallel observations. It is only in the case of classical states that these contrasting viewpoints amount to the same thing.

Within the present framework the vector of unary readouts is simply the initial *unary* state of the system and, by itself, it does not provide a full characterisation. What is sometimes referred to as a 'state' is actually a unary state and it does not determine future behaviour in the non-classical case because many different initial states are locally similar, that is have the same unary restriction, and yet yield different readouts from subsequent observation. The corresponding variability in one's description of what is going on comes from the incomplete specification of the initial state. This simple fact has often been obscured by philosophical speculation. It then leads to the thesis that, since the so-called 'state' does not determine the future, there is an inherent chance mechanism at work in the world which cannot be discussed within a deterministic framework. Be that as it may, and we shall not pursue the matter here, these issues are not substantive, they are questions about what sort of vocabulary is most fruitful for scientific discourse.

6. Temporal development

We now modify the static framework of the preceding sections by bringing dynamic features into explicit consideration. One can do so by taking some unary pointers to be clocks. In a non-relativistic setting, however, it is simpler to suppose that the state of an unobserved system changes with the passage of time in a well-defined way. This is the procedure adopted here. Throughout the section D_* is a simple system, $S \subseteq St$ is a designated subset of its states with $SD_* \subseteq S$. All the states mentioned are in S. For simplicity the temporal domain is taken to be the real line.

The temporal development of D_* in the absence of observation is specified by a set of transformations $\gamma_{a,b}: S \to S$, defined for real numbers $a \leqslant b$, such that $\gamma_{a,a}$ is the identity map and

$$\gamma_{a,b}\,\gamma_{b,c} = \gamma_{a,c},\quad a \leqslant b \leqslant c \tag{6.1}$$

If the state at time a+ is s and the system is unobserved up to time $b > a$, then the state at time b is $s\gamma_{a,b}$. If the system remains unobserved up to time $c > b$ the state is then $(s\gamma_{a,b})\gamma_{b,c}$, (6.1) ensures that this is $s\gamma_{a,c}$. The set $\Gamma_+ = \{\gamma_{a,b}: b \geqslant a\}$ is called the future schema. It is said to be stationary when, for all c and all $b \geqslant a$,

$$\gamma_{a+c,b+c} = \gamma_{a,b} = \gamma_{0,b-a} \ .$$

In that case we write Γ_t instead of $\gamma_{0,t}$ and (6.1) reads

$$\Gamma_t \Gamma_\tau = \Gamma_{t+\tau} \ .$$

The future schema $\Gamma_+ = \{\Gamma_t : t \geqslant 0\}$ is then a one-parameter semigroup of transformations of S. For simplicity of exposition we shall deal only with the stationary case and in addition, assume that time is reversible. In other words Γ_t is defined for all real t, $\Gamma_{-t} = \Gamma_t^{-1}$ is the inverse of Γ_t and $\Gamma = \{\Gamma_t : t \text{ a real number}\}$ is a group. It is called the dynamic group of D_*.

If D_* is in state s at time 0+ and we subsequently observe it with the unary pointer p in D, then the readout at that observation and the state just after it will depend on the time at which it was made. If it takes place at time $t > 0$ and there was no prior observation, the readout is $p(s\Gamma_t)$. But the new state $(s\Gamma_t)p$ is not necessarily the same as $(sp)\Gamma_t$. In the non-classical case $(s\Gamma_t)p\Gamma_\tau$ may not be the same state as $s\Gamma_{t+\tau}$, $\tau > 0$, because the system is disturbed when it is observed with p at time t. If there is no other observation between t and $t + \tau$, then the readout from the pointer q in D at time $t + \tau$ is $q\{(s\Gamma_t)p\Gamma_\tau\}$ and, in general, this is not the same as $q(s\Gamma_{t+\tau})$.

The set $\Omega(s) = \{s\Gamma_t : t \text{ a real number}\}$ is called the orbit of s. Because of the group property in Γ, each state in S lies on exactly one orbit. Temporal development in the absence of observation can be thought of as motion along an orbit. But an act of observation may knock it out of one orbit into another. Suppose D_* is in state s at time 0 and is first observed at $t > 0$ with dimension d. The state at time $t-$ is $s\Gamma_t$ on $\Omega(s)$ but the state at $t+$ is $s\Gamma_t d$ and this may not lie on $\Omega(s)$. The system will then move along the new orbit $\Omega(s\Gamma_t d)$ until the next observation.

7. State dynamics

Observationally induced orbital shifts are reminiscent of certain qualitative features of quantum mechanics. In this section we indicate some formal analogies between the temporal development of simple systems and those of quantum mechanics. A fuller discussion will be given elsewhere.

Let S, with $SD_* \subseteq S$, be a set of states of the simple system D_* and let Γ be its dynamic group. Let H be a Hilbert space of complex-valued functions on S. For each ψ in H, define ψ_t in H by $\psi_t(s) = \psi(s\Gamma_t)$. For each t we define G_t: $H \to H$ by decreeing that $\psi G_t = \psi_t$. The set $G = \{G_t\colon t \text{ a real number}\}$ is a one-parameter group of transformations of H. We suppose it has an infinitesimal generator

$$H = i \lim_{t\to 0}\{(G_t - G_0)/t\}.$$

It follows at once that

$$i\dot{G}_t = HG_t = G_tH$$

and hence that

$$i\dot{\psi}_t = H\psi_t.$$

We call this the Schrödinger equation in the absence of observation because of its formal equivalence to the quantum mechanical equation of that name when we interpret H as the Hamiltonian.

If q is a pointer in D we regard it as an operator on H through the equations $\psi q(s) = \psi(sq)$. It is often convenient to regard q as function of time by decreeing that

$$q_t = G_t q\, G_t^{-1}.$$

It then follows that

$$i\dot{q}_t = G_t[H,q]\,G_t^{-1}$$

where $[H,q] = (Hq - qH)$. Taking $t = 0$ and writing $\dot{q}$ instead of $\dot{q}_0$, we obtain

$$i\dot{q} = [H,q].$$

This is formally equivalent to the usual quantum mechanical equations of motion for an operator-observable.